A U D I T   G U I D E

Audit Sampling

MARCH 1, 2012

his edition of the AICPA Audit Guide *Audit Sampling*, which was originally issued in 2008, has been modified by
he AICPA staff to include certain changes necessary because of the issuance of authoritative pronouncements
nce the guide was originally issued. The appendix "Schedule of Changes Made to the Text From the Previous
dition" identifies all changes made in this edition of the guide. The changes do *not* include all those that might
e considered necessary if the guide was subjected to a comprehensive review and revision.

Copyright © 2012 by
American Institute of Certified Public Accountants, Inc.
New York, NY 10036-8775

All rights reserved. For information about the procedure for requesting permission to make copies of any part of this work, please e-mail copyright@aicpa.org with your request. Otherwise, requests should be written and mailed to the Permissions Department, AICPA, 220 Leigh Farm Road, Durham, NC 27707-8110.

1 2 3 4 5 6 7 8 9 0 AAP 1 9 8 7 6 5 4 3 2

ISBN 978-1-93735-088-8

Important Notice to Reader

This AICPA Audit Guide has been fully conformed to reflect the new standards resulting from the Clarity Project. This year's edition of the guide fully incorporates the clarified auditing standards into all guide content, so that auditors can further their understanding of the clarified auditing standards, as well as begin updating their audit methodologies, resources, and tools prior to the clarified auditing standards' effective date. Additionally, this approach gives auditors the opportunity to review and understand the changes made by their third-party audit methodology and resource providers, if applicable. The clarified auditing standards are effective for audits of financial statements for periods ending on or after December 15, 2012 (calendar year 2012 audits). Auditors should continue to use the 2008 edition of this guide until the clarified auditing standards become effective for the auditors' engagements.

Preface

About AICPA Audit and Accounting Guides

This AICPA Audit Guide was prepared under the supervision of the AICPA *Audit Sampling* Guide Task Force. The purpose of the guide is to assist auditors fulfill their responsibilities in designing and performing sampling in a financial statement audit conducted in accordance with generally accepted auditing standards (GAAS) as established by the AICPA Auditing Standards Board (ASB) (United States). GAAS established by the ASB are applicable to audits of nonissuers. *Nonissuers* are entities other than *issuers*, as defined by the Sarbanes-Oxley Act (SOX), or other entities that are required to be audited by a registered public accounting firm as prescribed by the rules of the Securities and Exchange Commission (SEC).

Auditing guidance included in an AICPA Audit Guide is recognized as an interpretive publication as defined in AU-C section 200, *Overall Objectives of the Independent Auditor and the Conduct of an Audit in Accordance with Generally Accepted Auditing Standards* (AICPA, *Professional Standards*). Interpretive publications are recommendations on the application of GAAS in specific circumstances, including engagements for entities in specialized industries.

An interpretive publication is issued under the authority of the ASB after all ASB members have been provided an opportunity to consider and comment on whether the proposed interpretive publication is consistent with GAAS. The members of the ASB have found the auditing guidance in this guide to be consistent with existing GAAS.

Although interpretive publications are not auditing standards, AU-C section 200 requires the auditor to consider applicable interpretive publications in planning and performing the audit because interpretive publications are relevant to the proper application of GAAS in specific circumstances. If the auditor does not apply the auditing guidance in an applicable interpretive publication, the auditor should document how the requirements of GAAS were complied with in the circumstances addressed by such auditing guidance.

Purpose and Applicability

This AICPA Audit Guide presents recommendations on the application of GAAS to audits involving the use of audit sampling methods. It is an update of the 2008 AICPA Audit Guide by the same name. The auditing guidance in this edition of the guide has been conformed to Statement on Auditing Standards (SAS) Nos. 122–125, which were issued in 2011 as part of the ASB's Clarity Project. The guide also includes increased guidance on the use of nonstatistical audit sampling. This guidance is more integrated and explains throughout the guide the common factors that need to be considered when following either a statistical or nonstatistical approach. Although the purpose of this guide is to provide guidance to help auditors apply audit sampling in accordance with AU-C section 530, *Audit Sampling* (AICPA, *Professional Standards*), the concepts and procedures described herein may be useful when performing attestation engagements that involve sampling.

Recognition

Darrel R. Schubert
Chair, Auditing Standards Board

2008 *Audit Sampling* Guide Task Force

Lynford Graham, *Chair*
Abraham D. Akresh
John P. Brolly
Michael F. Campana
Mark S. Chapin
William L. Felix, Jr.
Kenneth C. Garrett
Mark D. Mayberry
Douglas F. Prawitt
Trevor R. Stewart
John R. Troyer
Phil D. Wedemeyer
Harold I. Zeidman

AICPA Staff

Dennis W. Ridge, Jr.
Technical Manager
Accounting & Auditing Publications

The 2008 *Audit Sampling* Guide Task Force gratefully acknowledges the contributions of reviewers of an earlier draft of the guide, Douglas Prawitt and Steven Glover of Brigham Young University. Special recognition is given to Abraham Akresh who also served as a member of the 1983 and 2001 *Audit Sampling* Guide Task Forces. Appreciation is also extended to Donald M. Roberts for his insight and commentary.

The AICPA thanks Lynford Graham for his invaluable assistance in updating the guidance in the 2012 edition of the guide, including conforming the guide to the clarified auditing standards issued by the ASB.

The AICPA also thanks those who reviewed drafts of the conforming changes to this edition of the guide: Cheryl E. Clark, Collette Cummins, James A. Clous, Lucas Hoogduin, Robert Dohrer, Steven M. Glover, Maria Manasses, Mark D. Mayberry, Keith Newton, Tom Shepherd, Ann Thornton, and Harold I. Zeidman.

Guidance Considered in This Edition

This edition of the guide has been modified by the AICPA staff to include certain changes necessary due to the issuance of authoritative guidance since the guide was originally issued, and other revisions as deemed appropriate. Authoritative guidance issued through March 1, 2012, has been considered in the development of this edition of the guide.

Authoritative guidance that is issued and effective for entities with fiscal years ending on or before March 1, 2012, is incorporated directly in the text of this guide. The presentation of authoritative guidance issued but not yet effective as of March 1, 2012, for entities with fiscal years ending after that same date is being presented differently than in past editions of this guide. This information is being presented as a guidance update, which is a shaded area that contains information on the new guidance. The distinct presentation of this content is intended to aid the reader in differentiating content that may not be effective for the reader's purposes.

This guide includes relevant guidance issued up to and including the following:

- SAS No. 125, *Alert That Restricts the Use of the Auditor's Written Communication* (AICPA, *Professional Standards*, AU-C sec. 905)
- Interpretation No. 1, "Dating the Auditor's Report on Supplementary Information," of AU section 551, *Supplementary Information in Relation to the Financial Statements as a Whole* (AICPA, *Professional Standards*, AU sec. 9551 par. .01–.04)
- Revised interpretations issued through March 1, 2012, including Interpretation No. 1, "Use of Electronic Confirmations," of AU section 330, *The Confirmation Process* (AICPA, *Professional Standards*, AU sec. 9330 par. .01–.08)
- Statement of Position 09-1, *Performing Agreed-Upon Procedures Engagements That Address the Completeness, Accuracy, or Consistency of XBRL-Tagged Data* (AICPA, *Technical Practice Aids*, AUD sec. 14,440)

Users of this guide should consider guidance issued subsequent to those items listed previously to determine their effect on entities covered by this guide. In determining the applicability of recently issued guidance, its effective date should also be considered.

The changes made to this edition of the guide are identified in appendix H, "Schedule of Changes Made to the Text From the Previous Edition." The changes do not include all those that might be considered necessary if the guide were subjected to a comprehensive review and revision.

Defining Professional Responsibilities

AICPA *Professional Standards* for audit engagements use the following two categories of professional requirements, identified by specific terms, to describe the degree of responsibility it imposes on auditors:

- *Unconditional requirements.* The auditor must comply with an unconditional requirement in all cases in which such requirement is relevant. GAAS use the word *must* to indicate an unconditional requirement.
- *Presumptively mandatory requirements.* The auditor must comply with a presumptively mandatory requirement in all cases in which such a requirement is relevant except in rare circumstances. GAAS use the word *should* to indicate a presumptively mandatory requirement.

In rare circumstances, the auditor may judge it necessary to depart from a relevant presumptively mandatory requirement. In such circumstances, the auditor should perform alternative audit procedures to achieve the intent of that requirement. The need for the auditor to depart from a relevant presumptively mandatory requirement is expected to arise only when the requirement is for a specific procedure to be performed and, in the specific circumstances of the audit, that procedure would be ineffective in achieving the intent of the requirement.

Prior to SAS No. 122, the phrase *is required to* or *requires* was used to express an unconditional requirement in GAAS (equivalent to *must*). With the issuance of SAS No. 122, the phrases *is required to* and *requires* do not convey a requirement or the degree of responsibility it imposes on auditors. Instead those terms are used to express that a requirement exists. The terms are typically used in the clarified auditing standards to indicate that a requirement exists elsewhere in GAAS.

Applicability of GAAS and Public Company Accounting Oversight Board Standards

Audits of the financial statements of *nonissuers* (those entities not subject to the Sarbanes-Oxley Act of 2002 or the rules of the SEC—that is, private entities, generally speaking) are conducted in accordance with GAAS as issued by the ASB, the designated senior committee of the AICPA with the authority to promulgate auditing standards for nonissuers. The ASB develops and issues standards in the form of SASs through a due process that includes deliberation in meetings open to the public, public exposure of proposed SASs, and a formal vote. The SASs and their related interpretations are codified in the AICPA's *Professional Standards*. Rule 202, *Compliance With Standards* (AICPA, *Professional Standards*, ET sec. 202 par. .01), of the AICPA Code of Professional Conduct requires an AICPA member who performs an audit to comply with the standards promulgated by the ASB. Failure to follow GAAS, and any other applicable auditing standards, is a violation of that rule.

Audits of the financial statements of *issuers*, as defined by the SEC (those entities subject to SOX or the rules of the SEC—that is, public entities, generally speaking), are conducted in accordance with standards established by the Public Company Accounting Oversight Board (PCAOB), a private sector, nonprofit corporation created by SOX to oversee the audits of issuers. The SEC has oversight authority over the PCAOB, including the approval of its rules, standards, and budget.

References to Professional Standards

In citing GAAS and their related interpretations, references use section numbers within the codification of currently effective SASs and not the original statement number, as appropriate.

AICPA.org Website

The AICPA encourages you to visit the website at www.aicpa.org, and the new Financial Reporting Center at www.aicpa.org/FRC. The Financial Reporting Center was created to support members in the execution of high-quality financial reporting. Whether you are a financial statement preparer or a member in public practice, this center provides exclusive member-only resources for the entire financial reporting process, and provides timely and relevant news, guidance and examples supporting the financial reporting process, including accounting, preparing financial statements and performing compilation, review, audit, attest or assurance and advisory engagements. Certain content on the AICPA's websites referenced in this guide may be restricted to AICPA members only.

Select Recent Developments Significant to This Guide

ASB's Clarity Project

To address concerns over the clarity, length, and complexity of its standards, the ASB has made a significant effort to clarify the SASs. The ASB established clarity drafting conventions and undertook to redraft all of its SASs in accordance with those conventions, which include the following:

- Establishing objectives for each clarified SAS
- Including a definitions section, where relevant, in each clarified SAS
- Separating requirements from application and other explanatory material
- Numbering application and other explanatory material paragraphs using an A- prefix and presenting them in a separate section that follows the requirements section
- Using formatting techniques, such as bulleted lists, to enhance readability
- Including, when appropriate, special considerations relevant to audits of smaller, less complex entities within the text of the clarified SAS
- Including, when appropriate, special considerations relevant to audits of governmental entities within the text of the clarified SAS

In addition, as the ASB redrafted standards for clarity, it also converged the standards with the International Standards on Auditing (ISA), issued by the International Auditing and Assurance Standards Board. As part of redrafting the standards, they now specify more clearly the objectives of the auditor and the requirements which the auditor has to comply with when conducting an audit in accordance with GAAS.

With the release of SAS Nos. 117–120 and Nos. 122–125, the project is near completion. As of the date of this guide, the only SASs remaining to be clarified are

- SAS No. 59, *The Auditor's Consideration of an Entity's Ability to Continue as a Going Concern*, as amended (AICPA, *Professional Standards*, AU sec. 341) and
- SAS No. 65, *The Auditor's Consideration of the Internal Audit Function in an Audit of Financial Statements* (AICPA, *Professional Standards*, AU sec. 322).

Note that SAS No. 122 withdraws SAS No. 26, *Association With Financial Statements,* as amended (AICPA, *Professional Standards*), from professional standards.

AAG-SAM

SAS Nos. 122–125 will be effective for audits of financial statements for periods ending on or after December 15, 2012. Refer to individual AU-C sections for specific effective date language.

As part of the Clarity Project, current AU section numbers have been renumbered based on equivalent ISAs. Guidance is located in "AU-C" section numbers instead of "AU" section numbers. "AU-C" is a temporary identifier to avoid confusion with references to existing "AU" sections, which remain effective through 2013, in AICPA *Professional Standards*. The "AU-C" identifier will revert to "AU" in 2014, by which time the clarified auditing standards become fully effective for all engagements. Note that AU-C section numbers for clarified SASs with no equivalent ISAs have been assigned new numbers. The ASB believes that this recodification structure will aid firms and practitioners that use both ISAs and GAAS.

All auditing interpretations corresponding to a SAS have been considered in the development of a clarified SAS and incorporated accordingly, and have been withdrawn by the ASB except for certain interpretations that the ASB has retained and revised to reflect the issuance of SAS No. 122. The effective date of the revised interpretations aligns with the effective date of the corresponding clarified SAS.

Important Notice to Reader

This AICPA Audit Guide has been fully conformed to reflect the new standards resulting from the Clarity Project. This year's edition of the guide fully incorporates the clarified auditing standards into all guide content, so that auditors can further their understanding of the clarified auditing standards, as well as begin updating their audit methodologies, resources, and tools prior to the clarified auditing standards' effective date. Additionally, this approach gives auditors the opportunity to review and understand the changes made by their third-party audit methodology and resource providers, if applicable. The clarified auditing standards are effective for audits of financial statements for periods ending on or after December 15, 2012 (calendar year 2012 audits). Auditors should continue to use the 2008 edition of this guide until the clarified auditing standards become effective for the auditors' engagements.

See the previous section titled "Guidance Considered in this Edition" for more information related to the guidance issued as of the date of this guide. See also appendix F, "Mapping and Summarization of Changes—Clarified Auditing Standards." This appendix cross references extant AU sections with AU-C sections and indicates the nature of changes made in the clarified standard.

Applicability of Requirements of SOX

Publicly held companies and other *issuers* (see the following definition) are subject to the provisions of SOX and related SEC regulations implementing SOX. Their outside auditors are also subject to the provisions of SOX and to the rules and standards issued by the PCAOB.

Presented in the following paragraphs is a summary of certain key areas addressed by SOX, the SEC, and the PCAOB that are particularly relevant to the preparation and issuance of an issuer's financial statements and the preparation and issuance of an audit report on those financial statements. However, the provisions of SOX, the regulations of the SEC, and the rules and standards of the PCAOB are numerous and are not all addressed in this section or in this guide.

Definition of an *Issuer*

SOX states that the term *issuer* means an issuer (as defined in Section 3 of the Securities Exchange Act of 1934 (15 U.S.C. 78c)), the securities of which are registered under Section 12 of that act (15 U.S.C. 78l), or that is required to file reports under Section 15(d) (15 U.S.C. 78o(d)), or that files or has filed a registration statement that has not yet become effective under the Securities Act of 1933 (15 U.S.C. 77a et seq.), and that it has not withdrawn.

Issuers, as defined by SOX, and other entities when prescribed by the rules of the SEC (collectively referred to in this guide as *issuers* or *issuer*) and their public accounting firms (who must be registered with the PCAOB) are subject to the provisions of SOX, implementing SEC regulations, and the rules and standards of the PCAOB, as appropriate.

Nonissuers are those entities not subject to SOX or the rules of the SEC.

Introduction

> **© Update I-1 *Audit*: Clarified Auditing Standards**
>
> The auditing guidance in this guide edition has been conformed to Statement on Auditing Standards (SAS) Nos. 122–125, which were issued in 2011 as part of the Auditing Standards Board's (ASB's) Clarity Project. These clarified SASs are effective for audits of financial statements for periods ending on or after December 15, 2012. Although extensive, the revisions to generally accepted auditing standards (GAAS) resulting from these clarified SASs do not change many of the requirements found in the auditing standards that they supersede.
>
> To assist auditors and financial reporting professionals in making the transition, this guide includes appendix F, "Mapping and Summarization of Changes—Clarified Auditing Standards," which provides a cross reference of the sections in the superseded auditing standards to the applicable sections in the clarified auditing standards and identifies the changes, either substantive or primarily clarifying in nature, that may affect an auditor's practice or methodology relative to the applicable sections of SAS Nos. 122–125. It also summarizes the changes resulting from the requirements of SAS Nos. 122–125.
>
> The preface of this guide and the Financial Reporting Center on www.aicpa.org provide more information on the Clarity Project. Visit www.aicpa.org/sasclarity.

The Development of Audit Sampling

I.1 At the beginning of the twentieth century, the rapid increase in the size of American companies created a need for audits based on selected tests of items constituting account balances or classes of transactions. Previously, a number of audits had included an examination of every transaction in the period covered by the financial statements. At that time, professional literature paid little attention to the subject of sampling.

I.2 A program of audit procedures printed in 1917 in the *Federal Reserve Bulletin* included some early references to sampling, such as selecting "a few book items" of inventory. The program was prepared by a special committee of the AICPA's earliest predecessor, the American Association of Public Accountants.

I.3 For the first few decades of the twentieth century, auditors often applied sampling, but the extent of sampling was not related to the effectiveness of an entity's internal control. Some auditing articles and textbooks in the 1910s and 1920s referred to reducing the extent of tests of details based on reliance on the entity's internal check, as internal control was first called; however, there was little acceptance of this relationship in practice until the 1930s.

I.4 In 1955, the American Institute of Accountants (later to become the AICPA) published *A Case Study of the Extent of Audit Samples*, which summarized audit programs prepared by several CPAs to indicate the extent of audit sampling each considered necessary for a case study audit. The study was important because it was one of the first professional publications on audit sampling. It also acknowledged some relationship between the extent of tests of details and reliance on internal control. The 1955 study concluded, "Although there was some degree of similarity among the views expressed as to the extent of sampling necessary for most items in the financial statements, no clear-cut pattern resulted."

I.5 During the 1950s, some interest developed in applying statistical principles to sampling in auditing. Some auditors succeeded in developing methods for applying statistical sampling; however, other auditors questioned whether those techniques should be applied in auditing.

I.6 The first pronouncement on the subject of statistical sampling in auditing was the special report *Statistical Sampling and the Independent Auditor* issued by the AICPA's Committee on Statistical Sampling in 1962. The report concluded that statistical sampling was permitted under GAAS. A second report, *Relationship of Statistical Sampling to Generally Accepted Auditing Standards*, issued by the committee in 1964, illustrated the relationship between precision and confidence (reliability) in sampling and GAAS. The 1964 report was later included as appendix A of Statement on Auditing Procedure (SAP) No. 54, *The Auditor's Study and Evaluation of Internal Control*. The statement elaborated on the guidance provided by the earlier report. The Auditing Procedures Committee report *Precision and Reliability for Statistical Sampling in Auditing* was issued in 1972 as appendix B of SAP No. 54.

I.7 Two other SAPs included references to sampling applications in auditing. SAP No. 33, *Auditing Standards and Procedures (a codification)*, issued in 1963, indicated that a practitioner might consider using statistical sampling in appropriate circumstances. SAP No. 36, *Revision of "Extensions of Auditing Procedure" Relating to Inventories*, issued in 1966, provided guidance on the auditor's responsibility when a client uses a sampling procedure, rather than a complete physical count, to determine inventory balances.

I.8 From 1967 to 1974, the AICPA published a series of volumes on statistical sampling, *An Auditor's Approach to Statistical Sampling*, for use in continuing professional education. In 1978, the AICPA published *Statistical Auditing*, by Donald M. Roberts, explaining the theory underlying statistical sampling in auditing.

I.9 In 1981, the AICPA's ASB issued SAS No. 39, *Audit Sampling* (AICPA, *Professional Standards*, AU sec. 350), which provides general guidance on both nonstatistical and statistical sampling in auditing and superseded appendixes A and B of SAS No. 1, *Codification of Auditing Standards and Procedures* (AICPA, *Professional Standards*). In 1983, the AICPA published the first edition of this Audit Guide *Audit Sampling*. In 2001, the AICPA published an updated version of the guide.

I.10 In 2006, the ASB issued a suite of eight risk assessment standards (SAS Nos. 104–111) to be used in the planning and performance of a financial statement audit. Several of these pronouncements also provide guidance on the use of audit sampling. SAS No. 107, *Audit Risk and Materiality in Conducting an Audit* (AICPA, *Professional Standards*, AU sec. 312), established requirements and provided guidance on the auditor's consideration of audit risk and materiality when planning and performing an audit of financial statements in accordance with GAAS. Audit risk and materiality are important in determining the nature, timing, and extent of auditing procedures (including those that involve audit sampling) and evaluating the results of those procedures. SAS No. 109, *Understanding the Entity and Its Environment and Assessing the*

Risks of Material Misstatement (AICPA, *Professional Standards*, AU sec. 314), and SAS No. 110, *Performing Audit Procedures in Response to Assessed Risks and Evaluating the Audit Evidence Obtained* (AICPA, *Professional Standards*, AU sec. 318), clarify the circumstances under which controls can be relied on and the importance of IT general controls and tests of controls as a basis for reliance. The AICPA also issued the Audit Guide *Assessing and Responding to Audit Risk in a Financial Statement Audit* to provide guidance on obtaining an understanding of the entity and its environment, including its internal control, assessing the risks of material misstatement, designing further audit procedures that respond to the assessed risks, and evaluating audit findings and evidence. In discussing the auditor's assessment of control risk, the preceding guidance describes the manner in which the auditor designs, performs, and evaluates tests of controls, including those that involve audit sampling. In 2008, a significantly revised Audit Guide *Audit Sampling* was published by the AICPA.

I.11 In October 2011, the ASB issued SAS No. 122, *Statements on Auditing Standards: Clarification and Recodification* (AICPA, *Professional Standards*), as a result of the Clarity Project. SAS No. 122 recodified the AU section numbers in AICPA *Professional Standards* as designated by SAS Nos. 1–121. Clarified audit sampling requirements, along with other related requirements and guidance appearing in this guide, were carried forward by SAS No. 122 and recodified in AU-C section 530, *Audit Sampling*, and AU-C section 450, *Evaluation of Misstatements Identified in an Audit* (AICPA, *Professional Standards*), respectively.

The Significance of Audit Sampling

I.12 AU-C section 530 and AU-C section 330, *Performing Audit Procedures in Response to Assessed Risks and Evaluating the Audit Evidence Obtained* (AICPA, *Professional Standards*), recognize that auditors are often aware of items in account balances or classes of transactions that likely contain misstatements. Auditors consider this knowledge in planning procedures, including audit sampling. They usually will have no special knowledge about other items in account balances or classes of transactions that, in their judgment, will need to be tested to fulfill the audit objectives. Auditors might apply audit sampling to those account balances or classes of transactions. AU-C section 530 provides guidance for planning, performing, and evaluating audit samples using two approaches: nonstatistical and statistical.

The Purpose of This Guide

I.13 This guide provides guidance to help auditors apply audit sampling in accordance with AU-C section 530 and AU-C section 450. It provides practical guidance on the use of nonstatistical and statistical sampling in auditing. In many cases, auditors may apply procedures not involving audit sampling to account balances or classes of transactions. Neither this document nor AU-C section 530 establishes requirements or provides guidance on planning, performing, or evaluating audit procedures not involving audit sampling.[1]

---

[1] AU-C section 330, *Performing Audit Procedures in Response to Assessed Risks and Evaluating the Audit Evidence Obtained* (AICPA, *Professional Standards*), establishes requirements and provides guidance on planning, performing, or evaluating audit procedures not involving audit sampling.

I.14 This guide discusses several approaches to the application of sampling in auditing. It does not discuss the use of sampling if the objective of the application is to develop an original estimate of quantities or amounts. To avoid a complex, highly technical presentation, this guide does not include guidance on every possible valid method of selecting and evaluating audit samples. It also does not discuss the mathematical formulas underlying statistical sampling because knowledge of statistical sampling formulas, which was once required to apply statistical sampling in auditing, is no longer as important because the formulas are often imbedded in software that assists the auditor in sizing, selecting, and evaluating the sample. However, a reference document is available from the AICPA that illustrates the formulas underlying the various tables in this guide. "Technical Notes on the AICPA Audit Guide *Audit Sampling*"[2] contains key statistical formulas used to develop the tables in the guide for the benefit of statistical specialists, educators, students, and others. This paper may also help auditors extend the tables to cover parameters meeting firm-specific policies and guidance, and help individual practitioners tailor their sampling techniques to specific audit circumstances, and developers write software to augment or replace tables. This guide assumes that the auditor uses appropriate and reliable computer programs or tables to perform the calculations and selections necessary for statistical sampling.

I.15 This guide may be used both as a reference source for those who are knowledgeable about audit sampling and as initial background for those who are new to this area. Auditors unfamiliar with technical sampling considerations might benefit by combining use of this guide with a continuing education course in audit sampling and by consulting with persons knowledgeable in audit sampling. Training is available from several sources, including the AICPA, state CPA societies, colleges and universities, private vendors, and some CPA firms.

I.16 The guide is organized as follows:

- Chapter 1, "Characteristics of Audit Sampling," defines audit sampling and illustrates the difference between procedures that involve audit sampling and those that do not involve audit sampling.
- Chapter 2, "The Audit Sampling Process," provides overviews of the audit sampling process and the various approaches to audit sampling.
- Chapter 3, "Nonstatistical and Statistical Audit Sampling in Tests of Controls," provides guidance on the use of nonstatistical and statistical audit sampling for tests of controls.
- Chapter 4, "Nonstatistical and Statistical Audit Sampling for Substantive Tests of Details," provides general guidance on the use of nonstatistical and statistical audit sampling for substantive tests.
- Chapter 5, "Nonstatistical Sampling Case Study," provides a case study for nonstatistical sampling applications for substantive tests.
- Chapter 6, "Monetary Unit Sampling," discusses monetary unit sampling.

---

[2] This document is available for download from the AICPA website at www.aicpa.org/Publications/AccountingAuditing/KeyTopics/DownloadableDocuments/Sampling_Guide_Technical_Notes.pdf.

- Chapter 7, "Classical Variables Sampling," discusses classical variables sampling techniques using computer programs.
- Chapters 6–7 each include a case study illustrating the application of the guidance.
- This guide includes several appendixes. Appendix A, "Attributes Statistical Sampling Tables"; appendix B, "Sequential Sampling for Tests of Controls"; and appendix C, "Monetary Unit Sampling Tables," are useful primarily in applying certain statistical sampling approaches. Appendix D, "Ratio of Desired Allowance for Sampling Risk of Incorrect Rejection to Tolerable Misstatement," describes an approach to controlling the risk of incorrect rejection when planning an audit sampling application. Appendix E, "Multilocation Sampling Considerations," contains a discussion relating to designing samples for multilocation sampling. Appendix F, "Mapping and Summarization of Changes—Clarified Auditing Standards," provides a mapping of the extant AU section to the clarified AU-C sections. Also included are appendix G, "Glossary," and appendix H, "Schedule of Changes Made to the Text From the Previous Edition."

I.17 An auditor using nonstatistical sampling is not required to compute the sample size for the nonstatistical sampling application using statistical theory; however, paragraph .A14 of AU-C section 530, which clarifies that sample sizes of statistical and nonstatistical samples ordinarily would be comparable when the same sampling parameters are used, states the following:

> The decision whether to use a statistical or nonstatistical sampling approach is a matter for the auditor's professional judgment; however, sample size is not a valid criterion to use in deciding between statistical and nonstatistical approaches. An auditor who applies statistical sampling may use tables or formulas to compute sample size based on the factors in paragraph .A13. An auditor who applies nonstatistical sampling exercises professional judgment to relate the same factors used in statistical sampling in determining the appropriate sample size. Ordinarily, this would result in a sample size comparable with the sample size resulting from an efficient and effectively designed statistical sample, considering the same sampling parameters. This guidance does not suggest that the auditor using nonstatistical sampling also compute a corresponding sample size using an appropriate statistical technique.

I.18 This guide provides several quantitative illustrations of sample sizes based on statistical theory that may be helpful to an auditor applying professional judgment and experience in considering the effect of various planning considerations on sample size when using nonstatistical sampling.[3]

I.19 When using audit sampling, the auditor chooses between a statistical and a nonstatistical approach to audit sampling. Both methods comply with auditing standards. Statistical methods are drawn from the field of

---

[3] Even though sample sizes between statistical and nonstatistical samples may be similar, other characteristics of the sampling plan such as sample selection methods may not be similar. Further adjustments to the nonstatistical sample plan, for example an increase in the sample size or changes in the selection method, may be needed to provide equivalent assurance from statistical and nonstatistical sampling plans.

applied statistics and require training and experience in their use. Nonstatistical methods draw on the auditor's experience and professional judgment in selecting items for evidence from populations and evaluating the results. In using statistical sampling, the auditor uses experience and judgment when determining the appropriate selection and evaluation methods provided from the field of applied statistics. It is important to note that nonstatistical sampling methods may use tools from statistical sampling such as random selection of sample items or determining sample size by using statistical sampling tables. A distinguishing element is the evaluation method where statistical methods state a specific numerical sampling risk in inferring the condition of the population from the sample. The differences between these two methods include the different levels of formality in structuring the design and execution of the procedures and the numerical control of and evaluation of sampling risk provided by statistical methods. Both approaches are best carried out by auditors who have training in their use and evaluation. Training in nonstatistical sampling generally provides an overview of statistical principles, because those principles are useful in helping the auditor to understand nonstatistical sampling.

I.20 Although the purpose of this guide is to provide guidance to help auditors apply audit sampling in accordance with AU-C section 530 and AU-C section 450, the concepts and procedures are useful when performing attestation engagements that involve audit sampling.

Audit Sampling Guidance for Compliance Audits

I.21 This guide is also applicable to the financial statement portion of audits conducted in accordance with the Single Audit Act and Office of Management and Budget (OMB) Circular A-133, *Audits of States, Local Governments, and Non-Profit Organizations*.

I.22 Chapter 11, "Audit Sampling Considerations of Circular A-133 Compliance Audits," in the AICPA Audit Guide Government Auditing Standards *and Circular A-133 Audits*, provides audit sampling guidance specific to compliance audits performed under OMB Circular A-133 requirements. This additional guidance is focused only on the compliance controls and compliance testing requirements of OMB Circular A-133 and is not applicable to the audits of financial statements of entities subject to requirements of the Single Audit Act. However, in some instances there will be samples that can provide evidence for both compliance and financial statement audit purposes, and it is important that the auditor use his or her professional judgment when determining whether samples meet the requirements for both purposes. For example, OMB Circular A-133 guidance defines three specific levels of compliance controls and compliance test sample sizes for use, depending on certain factors. There is no such guidance for audits of the financial statements. As another example, compliance test samples are generally limited to 60 items for most compliance tests under OMB Circular A-133, but no such limit exists for samples performed to gather evidence supporting the fairness of the financial statements. Portions of this guide are referenced in the AICPA Audit Guide Government Auditing Standards *and Circular A-133 Audits*. Other portions of this guide may also be helpful when performing audit sampling in compliance audits.

TABLE OF CONTENTS

Chapter		Paragraph
1	**Characteristics of Audit Sampling**	.01-.29
	Audit Sampling Defined	.04-.05
	Procedures That May Not Involve Audit Sampling	.06-.20
	Inquiry and Observation	.07
	Analytical Procedures	.08-.09
	Procedures Applied to Every Item in a Population	.10-.12
	Some Tests of Controls May Not Involve Audit Sampling	.13-.14
	Tests of Controls When Extrapolation is Not Intended	.15
	Procedures That Do Not Evaluate Characteristics	.16-.17
	Untested Balances	.18
	Tests of Automated IT Controls	.19-.20
	Sampling and Nonsampling Audit Procedures Distinguished	.21-.25
	Terminology Used in This Guide	.26-.29
	Reliability or Confidence Level	.27
	Sampling Risk	.28
	Precision	.29
2	**The Audit Sampling Process**	.01-.55
	Purpose and Nature of Audit Sampling	.02
	How Audit Sampling Differs From Sampling in Other Professions	.03-.06
	Evaluation of Audit Samples	.07
	Types of Audit Tests	.08-.14
	Tests of Controls	.09-.10
	Substantive Procedures	.11
	Dual-Purpose Tests	.12-.14
	Risk	.15-.21
	Sampling Risk	.19
	Nonsampling Risk	.20-.21
	Nonstatistical and Statistical Sampling	.22-.29
	Planning the Audit Sampling Procedures	.30-.34
	Types of Statistical Sampling Plans	.35-.43
	Attributes Sampling	.35-.37
	Variables Sampling	.38-.41
	Relating Balance Sheet and Income Statement Sampling	.42-.43
	General Implementation Considerations	.44-.55
	Continuing Professional Education	.45-.48
	Sampling Guidelines	.49
	Use of Specialists	.50-.51
	Supervision and Review	.52-.55

Chapter		Paragraph
3	Nonstatistical and Statistical Audit Sampling in Tests of Controls	.01-.98
	Determining the Test Objectives	.02-.05
	Defining the Deviation Conditions	.06
	Defining the Population	.07-.10
	Defining the Period Covered by the Test	.11-.21
	Initial Testing	.14
	Estimating Population Characteristics	.15-.17
	Considering the Completeness of the Population	.18-.21
	Defining the Sampling Unit	.22-.24
	The Role of Walkthroughs	.25-.28
	Determining the Method of Selecting the Sample	.29-.36
	Simple Random Sampling	.30
	Systematic Sampling	.31-.32
	Haphazard Sampling	.33-.34
	Block Sampling	.35-.36
	Determining the Sample Size	.37-.65
	Considering Sampling Risk in Assessing Controls Effectiveness	.38-.45
	Considering Other Evidence in Determining Risk of Concluding Controls are More Effective Than They Actually Are (Overreliance) and Tolerable Rate of Deviation	.46
	Considering the Risk of Concluding Controls are More Effective Than They Actually Are (Overreliance) for Multiple Controls Addressing the Same Control Objective	.47
	Determining the Tolerable Rate of Deviation	.48-.54
	Considering the Expected Population Deviation Rate	.55-.58
	Considering the Effect of Population Size	.59-.61
	Small Populations and Infrequently Operating Controls	.62-.63
	Considering a Sequential or a Fixed Sample Size Approach	.64
	Developing Sample Size Guidelines	.65
	Performing the Sampling Plan	.66-.72
	Voided Documents	.67
	Unused or Inapplicable Documents	.68
	Mistakes in Estimating Population Sequences	.69-.70
	Stopping the Test Before Completion	.71
	Inability to Examine Selected Items	.72
	Evaluating the Sample Results	.73-.95
	Calculating the Deviation Rate	.74
	Considering Sampling Risk	.75-.79
	Considering the Qualitative Aspects of the Deviations	.80-.81
	Extending the Sample When Control Deviations are Found	.82-.84

Chapter		Paragraph
3	**Nonstatistical and Statistical Audit Sampling in Tests of Controls—continued**	
	Assessing the Potential Magnitude of a Control Deficiency	.85-.94
	Reaching an Overall Conclusion	.95
	Documenting the Sampling Procedure	.96-.98
4	**Nonstatistical and Statistical Audit Sampling for Substantive Tests of Details**	.01-.108
	Determining the Test Objectives	.04-.05
	Defining the Population	.06-.12
	Considering the Completeness of the Population	.08-.10
	Identifying Individually Significant Items	.11-.12
	Defining the Sampling Unit	.13-.14
	Choosing an Audit Sampling Technique	.15-.16
	Selecting the Sample	.17-.22
	Determining the Sample Size	.23-.74
	Considering Variation Within the Population	.27-.32
	Determining the Acceptable Level of Risk	.33-.47
	Considering Tolerable Misstatement	.48-.49
	Performance Materiality and Tolerable Misstatement	.50-.59
	Considering the Expected Amount of Misstatement	.60-.61
	Considering the Effect of Population Size	.62
	Relating the Factors to Determine the Sample Size	.63-.74
	Performing the Sampling Plan	.75-.76
	Evaluating the Sample Results	.77-.104
	Projecting the Misstatement to the Population	.77-.89
	The Sufficiency of Sampling Evidence for Proposing Adjustments	.90
	Negative Confirmations	.91
	Interim Sample Results	.92
	Considering Sampling Risk at the Test Level	.93-.100
	Misstatements Not Projected	.101-.104
	Documenting the Sampling Procedure	.105-.108
5	**Nonstatistical Sampling Case Study**	.01-.16
	Determining the Sample Size	.08-.11
	Evaluating the Sample Results	.12-.16
6	**Monetary Unit Sampling**	.01-.63
	Selecting a Statistical Approach	.04-.08
	Advantages	.05-.06
	Disadvantages	.07-.08
	Defining the Sampling Unit	.09-.10
	Selecting the Sample	.11-.19

Chapter		Paragraph
6	**Monetary Unit Sampling—continued**	
	Determining the Sample Size	.20-.31
	Formula Method—No Misstatements Expected	.23-.25
	Formula Method—Some Misstatements Expected	.26-.31
	Evaluating the Sample Results	.32-.52
	Sample Evaluation With 100 Percent Misstatements	.35-.41
	Sample Evaluation With Less Than 100 Percent Misstatements	.42-.48
	Quantitative Considerations	.49-.51
	Qualitative Considerations	.52
	MUS Sampling Case Study	.53-.63
	Selecting the Sample	.56-.58
	Evaluating the Sample Results	.59-.63
7	**Classical Variables Sampling**	.01-.48
	Selecting a Statistical Approach	.03-.06
	Advantages	.04
	Disadvantages	.05-.06
	Types of Classical Variables Sampling Techniques	.07-.10
	Mean-Per-Unit Approach	.08
	Difference Approach	.09
	Ratio Approach	.10
	Choosing a Classical Variables Sampling Approach	.11-.15
	The Ability to Design a Stratified Sample	.12
	The Expected Number of Differences Between the Audited and Recorded Amounts	.13
	Required Information	.14-.15
	Determining the Sample Size	.16-.24
	Considering Variation Within the Population	.17-.19
	Calculating the Sample Size	.20-.24
	Evaluating the Sample Results	.25-.39
	Classical Variables Sampling Case Study	.40-.48

Appendix	
A	Attributes Statistical Sampling Tables
B	Sequential Sampling for Tests of Controls
C	Monetary Unit Sampling Tables
D	Ratio of Desired Allowance for Sampling Risk of Incorrect Rejection to Tolerable Misstatement
E	Multilocation Sampling Considerations
F	Mapping and Summarization of Changes—Clarified Auditing Standards
G	Glossary
H	Schedule of Changes Made to the Text From the Previous Edition
Index	

Chapter 1

Characteristics of Audit Sampling

> **⊙ Update 1-1 *Audit*: Clarified Auditing Standards**
>
> The auditing guidance in this guide edition has been conformed to Statement on Auditing Standards (SAS) Nos. 122–125, which were issued in 2011 as part of the Auditing Standards Board's Clarity Project. These clarified SASs are effective for audits of financial statements for periods ending on or after December 15, 2012. Although extensive, the revisions to generally accepted auditing standards resulting from these clarified SASs do not change many of the requirements found in the auditing standards that they supersede.
>
> To assist auditors and financial reporting professionals in making the transition, this guide includes appendix F, "Mapping and Summarization of Changes—Clarified Auditing Standards," which provides a cross reference of the sections in the superseded auditing standards to the applicable sections in the clarified auditing standards and identifies the changes, either substantive or primarily clarifying in nature, that may affect an auditor's practice or methodology relative to the applicable sections of SAS Nos. 122–125. It also summarizes the changes resulting from the requirements of SAS Nos. 122–125.
>
> The preface of this guide and the Financial Reporting Center on www.aicpa.org provide more information on the Clarity Project. Visit www.aicpa.org/sasclarity.

1.01 This chapter defines audit sampling and illustrates the difference between procedures that involve audit sampling and those that do not involve audit sampling.

1.02 An auditor often does not rely solely on the results of a single procedure to reach a conclusion on an assertion relating to an account balance or a class of transactions, or the operating effectiveness of controls. Rather, audit conclusions are usually based on evidence obtained from several sources as a result of applying a number of procedures. The combined evidence obtained from the various procedures is considered in reaching an opinion about whether the financial statements are free of material misstatement.

1.03 The assertions described in paragraph .A114 of AU-C section 315, *Understanding the Entity and Its Environment and Assessing the Risks of Material Misstatement* (AICPA, *Professional Standards*), should be considered when planning audit sampling (for example, what could go wrong or the correct population for sampling) as well as other audit procedures. In this guide, the guidance relating to balances and classes of transactions implies the consideration of relevant assertions for the particular account or class of transactions.

Observations and Suggestions

When indicating a best practice or providing guidance on the application of sampling procedures, this guide may use the terms *typically*, *normally*, *usually*, or *best practice*. These terms do not imply a requirement, but are

AAG-SAM 1.03

suggestions to assist auditors in identifying the usual circumstance or application of a concept.

Audit Sampling Defined

1.04 According to paragraph .05 of AU-C section 530, *Audit Sampling* (AICPA, *Professional Standards*), *audit sampling* is "The selection and evaluation of less than 100 percent of the population of audit relevance such that the auditor expects the items selected (the sample) to be representative[1] of the population and, thus, likely to provide a reasonable basis for conclusions about the population." In other words, audit sampling provides the auditor an appropriate basis on which to conclude on a characteristic of a population based on examining evidence regarding that characteristic from a sample of the population. Procedures not involving audit sampling are not the subject of AU-C section 530 or this guide.

1.05 In many contexts in sampling, "representative" conveys the sense that the sample results are believed to correspond, at the stated risk level, to what would have been obtained had the auditor examined all items in the population in the same way as examined in the sample. *Correspond* does not mean that the projected misstatement from the sample will exactly equal the misstatement in the population (which the auditor does not know). Rather a sample is expected to be representative if it is free from selection bias. Statistical samples are designed to be representative, with the stated confidence that the true population misstatement is measured by the confidence interval. Nonstatistical samples generally are selected in a way that the auditor expects them to be representative. Representative relates to the total sample, not to individual items in the sample. Also, representative does not relate to the sample size, but to how the sample was selected. The sample generally is expected to be representative only with respect to the occurrence rate or incidence of misstatements, not their specific nature. A sample misstatement due to an unusual circumstance may nevertheless be indicative of other unusual misstatements in the population.

Procedures That May Not Involve Audit Sampling

1.06 Some auditing procedures by their nature may not involve audit sampling (unless the procedures are specifically designed as audit samples). In general, procedures that may not involve audit sampling may be grouped into the categories as discussed in the following paragraphs.

Inquiry and Observation

1.07 Auditors ask many questions during the course of their audits. Auditors also observe the operations of their clients' businesses and their controls. Both inquiry and observation provide auditors with audit evidence. Inquiry and observation commonly are used in the following procedures:

- Interviewing management and employees
- Obtaining an understanding of the internal controls

---

[1] Appendix G, "Glossary," contains further discussion regarding the term *representative* in the context of audit sampling.

- Observing the behavior of personnel and the functioning of business operations
- Observing cash-handling activities
- Observing the operation of controls
- Performing walkthrough procedures[2]
- Observing the existence of land and buildings
- Obtaining written representations from management

In some cases these procedures could be designed as sampling procedures, such as designing multiple observations of physical security controls.

Analytical Procedures

1.08 According to paragraph .04 of AU-C section 520, *Analytical Procedures* (AICPA, *Professional Standards*), analytical procedures are defined as

> evaluations of financial information through analysis of plausible relationships among both financial and nonfinancial data. Analytical procedures also encompass such investigation, as is necessary, of identified fluctuations or relationships that are inconsistent with other relevant information or that differ from expected values by a significant amount.

In performing analytical procedures, the auditor "compares the recorded amounts or ratios developed from recorded amounts with expectations" developed by the auditor.

1.09 These procedures are not considered audit sampling because they do not result in projecting the result of the examination of a portion of the population to the total population. For similar reasons, scanning accounting records for unusual items is not audit sampling.

Procedures Applied to Every Item in a Population

1.10 In some circumstances, an auditor might decide to examine every item constituting an account balance or a class of transactions. Because the auditor is examining the entire population, rather than only a portion, to reach a conclusion about the balance or class as a whole, 100 percent examination is not a procedure that involves audit sampling. In some cases, the use of computer assisted audit techniques may allow the application of a test to all items in the population (for example, tests of clerical accuracy and comparison of invoices and shipments) and, thus, audit sampling does not apply.

1.11 A population for audit sampling purposes does not necessarily need to be an entire account balance or class of transactions. In some circumstances, an auditor might examine all the items that constitute an account balance or class of transactions that exceed a given amount (for example, more than $25,000) or that have an unusual characteristic (for example, require dual signature approval for payment). The auditor might either (*a*) apply other auditing procedures (for example, targeted analytical procedures performed at a detailed level such as at the line-item or location level) to items that do not exceed that given amount or possess the unusual characteristic or (*b*) apply no detailed auditing procedures to them because there is an acceptably low *risk*

---

[2] Walkthroughs may also include an examination of evidence and reperformance, depending on their design and performance.

of material misstatement existing in the remaining items. Again, the auditor is not using audit sampling when applying procedures in this manner. Rather, the auditor has segregated the account or class of transactions into two groups. One group is tested 100 percent; the other group is tested by analytical or other auditing procedures or remains untested based on the low level of *risk of material misstatement* in the portion not subjected to 100 percent testing.

1.12 For the same reason, cutoff tests often do not involve audit sampling applications. In performing cutoff tests, auditors often examine all significant transactions for a sufficient period surrounding the cutoff date and, as a result, such tests often do not involve the application of audit sampling. However, one could design cutoff tests by using audit sampling when the volume of transactions during the period of interest is high.

Some Tests of Controls May Not Involve Audit Sampling

1.13 Auditors choose from a variety of methods, including inquiry, observation, inspection of documentary evidence, and reperformance, in evaluating the implementation of controls. Although many procedures where documentary evidence is examined or where the auditor reperforms a control involve audit sampling, many of the other methods may not involve sampling. Certain types of tests of controls, because of the nature of the procedures used, do not normally involve audit sampling. For example, tests of automated application controls are generally tested only once or a few times when effective IT general controls are present, and thus do not rely on the concepts of risk and tolerable deviation as applied in other sampling procedures. Sampling generally is not applicable to analyses of controls for determining the appropriate segregation of duties or other analyses that do not examine documentary evidence of performance. In addition, sampling may not apply to tests of certain documented controls or to analyses of the effectiveness of security and access controls. Sampling also may not apply to some tests directed toward obtaining audit evidence about the operation of the control environment or the accounting system, for example, inquiry or observation of explanation of variances from budgets when the auditor does not desire to estimate the rate of deviation from the prescribed control, or when examining the actions of those charged with governance for assessing their effectiveness.

1.14 In addition, when the performance of a control is not documented or evidenced, such as the performance of an automated control where no record of the control performance is retained, the concept of sampling such a control in the conventional sense may not be meaningful. For example, such a test may be performed contemporaneously with its occurrence or tested with a *test deck* of data with known properties that are designed to test the automated controls, and the extent of testing and the periods included in the test are determined based on the quality of the related IT general controls. Such tests often do not involve audit sampling.

Tests of Controls When Extrapolation is Not Intended

1.15 Observation of a client's physical inventory count activities is a test usually performed primarily through the auditor's observation of the operation of controls over inventory movement, counting procedures, and other activities used by the client to control the count of the inventory. The auditor's test counts of client counts may not be for extrapolating results, but may be for determining the adequacy and accuracy of the count procedures. Nevertheless,

Characteristics of Audit Sampling

the auditor considers the deviations and misstatements found. As such, when discrepancies in the count are identified, an assessment is made of the reasons for the discrepancy, and a recount may be indicated for some or all of the inventory items by a count team or in a location until the auditor is satisfied that the count is accurate. Using this procedure during the count may not involve the application of audit sampling. Even when extrapolation is not intended, the auditor still considers issues such as the extent of procedures performed and the possibility of bias in the selection of sample items.

Procedures That Do Not Evaluate Characteristics

1.16 Procedures from which the auditor does not intend to extend the resulting conclusion to the remaining items in the account balance or class of transactions do not require audit sampling. The auditor does not use audit sampling when he or she applies an auditing procedure to less than 100 percent of the items in an account balance or class of transactions as something other than evaluating a trait of the entire balance or class. For example, an auditor might trace several transactions through an entity's accounting system to obtain an understanding of the design of the entity's internal control. In such cases, the auditor's intent is to gain a general understanding of the accounting system or other relevant parts of the internal control, rather than to evaluate a characteristic of all transactions processed. As a result, the auditor may not be using audit sampling.

1.17 Occasionally, auditors perform such procedures as checking arithmetical calculations or tracing journal entries into ledger accounts on less than a 100 percent (test) basis. When such procedures are applied to less than 100 percent of the arithmetical calculations or ledger postings that affect the financial statements, audit sampling may not be involved if the procedure is not a test to evaluate a characteristic of an account balance or class of transactions, but is intended to provide only limited evidence that supplements the auditor's other audit evidence regarding a financial statement assertion or is designed to provide evidence only about the items tested.

Untested Balances

1.18 The auditor might decide that he or she need not apply any detailed audit procedures to an account balance or class of transactions if the auditor believes that there is an acceptably low *risk of material misstatement* existing in the account or class. Audit sampling is not relevant to untested balances.

Tests of Automated IT Controls

1.19 IT systems process transactions and other information consistently unless the systems or programs (or related tables, parameters, or similar items that affect how the programs process the data) are changed. Therefore, when testing the operations of automated controls, the auditor may adopt the strategy of testing one or a few of each type of transaction at a point in time and test general controls (for example, controls over implementation and changes to systems and programs, access and security, and computer operations) to provide evidence that the automated controls have been operating effectively over the audit period. When IT general controls are tested and determined to be effective, a single test of an automated control for each type of control operation may be sufficient to place reliance on the automated control during the period of the audit examination.

AAG-SAM 1.19

1.20 Because distinguishing between audit procedures involving audit sampling and procedures not involving audit sampling might be difficult, the next section of this chapter discusses the distinction between procedures that do and do not involve audit sampling.

Sampling and Nonsampling Audit Procedures Distinguished

1.21 An account balance or class of transactions may be examined by a combination of several audit procedures. These procedures might involve audit sampling. An illustration can help clarify the distinction between procedures that do or do not involve audit sampling. An auditor might be examining fixed asset additions of $2 million. These might include 5 additions totaling $1.6 million related to a plant expansion program and 400 smaller additions constituting the remaining $400,000 recorded amount. The auditor might decide that the 5 large additions are individually significant and need to be examined 100 percent and might then consider whether to apply audit sampling to the remaining 400 items. This decision is based on the auditor's determination of tolerable misstatement for the sample and the assessment of the *risks of material misstatement* in the $400,000, not on the percentage of the $2 million individually examined (in this case, 80 percent). Several possible approaches are discussed in the following 3 situations.

1.22 *Situation 1.* The auditor has performed other procedures related to fixed-asset additions, including the following:

- Risk assessment procedures
- The consideration of related controls, which supported a low level of assessed control risk
- A review of the entries in the fixed asset ledger, which revealed no unusual items
- An analytical procedure, which suggested the $400,000 recorded amount, does not contain a material misstatement

1.23 In this situation, the auditor might decide that sufficient audit evidence regarding fixed-asset additions has been obtained without applying audit sampling to the remaining individually insignificant items. Therefore, the concept of audit sampling would not apply unless a sample is selected.

1.24 *Situation 2.* The auditor has not performed any procedures related to the accuracy of the remaining 400 items, but, nonetheless, decides that any misstatement in those items would be immaterial. The physical existence of the assets was verified by other procedures. The only remaining exposure is assessed to be the *risks of material misstatement* in the accuracy of the recorded amounts, which, based on the simple cash based purchases and controls over disbursements, the auditor has assessed to be low. Therefore, the concept of audit sampling would not apply unless a sample is selected.

1.25 *Situation 3.* The auditor has performed some or all of the same procedures as in situation 1, but concludes that some additional audit evidence about the 400 individually insignificant additions will be obtained through audit sampling. In this case, the information in AU-C section 530 and this guide assists the auditor in planning, performing, and evaluating the audit sampling application.

Terminology Used in This Guide

1.26 The terms used in this guide are consistent with those in AU-C section 530 and other professional standards. Some auditors may be familiar with other terms, including *precision, confidence level, reliability, alpha risk,* and *beta risk*, which are often used in discussions of statistical sampling. AU-C section 530 does not use those terms because it applies to both statistical and nonstatistical sampling and, therefore, nontechnical terms are more appropriate. Also, certain statistical terms, such as *reliability* and *precision*, have been used with different meanings. Auditors may use various terms in their practice, as long as they understand the relationship of those terms to the concepts in AU-C section 530 and this guide. Terms used in this guide or found in various auditing literature are defined in the glossary found in appendix G, "Glossary." Some of those relationships follow.

Reliability or Confidence Level

1.27 AU-C section 530 and AU-C section 200, *Overall Objectives of the Independent Auditor and the Conduct of an Audit in Accordance With Generally Accepted Auditing Standards* (AICPA, *Professional Standards*), use the concept of *risk* instead of reliability (or confidence level). However, statistical sampling literature often uses the terms *reliability* and *confidence level*. In addition, other auditing standards use the term *assurance*, a concept related to confidence or reliability. Additionally, some auditors express the sampling guidance in their audit approaches in terms of *assurance,* and not *risk*. Risk is the complement of reliability or confidence level. For example, if an auditor accepts a 10 percent sampling risk, the reliability or confidence level is specified as 90 percent. The term *risk* is more consistent with the auditing framework described in the SASs. Audit professionals are advised to be familiar with the various terms that are relevant to audit sampling.

Sampling Risk

1.28 Paragraph .05 of AU-C section 530 defines sampling risk in terms of two types of erroneous conclusions:

    a. In the case of a test of controls, that controls are more effective than they actually are, or in the case of a test of details, that a material misstatement does not exist when, in fact, it does. The auditor is primarily concerned with this type of erroneous conclusion because it affects audit effectiveness and is more likely to lead to an inappropriate audit opinion.[3]

    b. In the case of a test of controls, that controls are less effective than they actually are, or in the case of a test of details, that a material misstatement exists when, in fact, it does not. This type of erroneous conclusion affects audit efficiency because it would usually lead to additional work to establish that initial conclusions were incorrect.[4]

---

[3] AU section 350, *Audit Sampling* (AICPA, *Professional Standards*), used the specific terms *risk of assessing control risk too low* (when sampling for tests of controls) and *risk of incorrect acceptance* (for substantive testing).

[4] AU section 350 used the specific terms *risk of assessing control risk too high* (controls) and *risk of incorrect rejection* (substantive).

Other sampling literature and paragraph .A13 in the "Application and Other Explanatory Material" section of AU-C section 530 term the risks in preceding subparagraph *a* as the *risk of overreliance* (for controls) and the *risk of incorrect acceptance* (for substantive testing). Formal statistical literature often terms this risk as *beta risk*. The risks described in preceding subparagraph *b* are also termed in prior AICPA and other sampling literature as the *risk of underreliance* (for controls) *and the risk of incorrect rejection* (for substantive tests). Formal statistical literature often terms this risk as *alpha risk*. Both *alpha risk* and *beta risk* (sometimes referred to as risks of type I and type II errors) are statistical terms that have not always been consistently applied in the auditing literature.

Precision

1.29 Precision might be used both as a planning concept and an evaluation concept for audit sampling. Rather than the term *precision*, AU-C section 530 uses the difference between the expected deviation rate or expected misstatement amount and the tolerable deviation rate or tolerable misstatement as a measure of precision.[5]

---

[5] This edition of the guide, as well as prior editions, use the term *allowance for sampling risk* to represent precision. Precision is a term used in statistical sampling.

Chapter 2

The Audit Sampling Process

> **⊕ Update 2-1 *Audit*: Clarified Auditing Standards**
>
> The auditing guidance in this guide edition has been conformed to Statement on Auditing Standards (SAS) Nos. 122–125, which were issued in 2011 as part of the Auditing Standards Board's Clarity Project. These clarified SASs are effective for audits of financial statements for periods ending on or after December 15, 2012. Although extensive, the revisions to generally accepted auditing standards resulting from these clarified SASs do not change many of the requirements found in the auditing standards that they supersede.
>
> To assist auditors and financial reporting professionals in making the transition, this guide includes appendix F, "Mapping and Summarization of Changes—Clarified Auditing Standards," which provides a cross reference of the sections in the superseded auditing standards to the applicable sections in the clarified auditing standards and identifies the changes, either substantive or primarily clarifying in nature, that may affect an auditor's practice or methodology relative to the applicable sections of SAS Nos. 122–125. It also summarizes the changes resulting from the requirements of SAS Nos. 122–125.
>
> The preface of this guide and the Financial Reporting Center on www.aicpa.org provide more information on the Clarity Project. Visit www.aicpa.org/sasclarity.

2.01 Audit sampling may be applied using statistical or nonstatistical approaches. This chapter provides overviews of the audit sampling process and the various approaches to audit sampling.

Purpose and Nature of Audit Sampling

2.02 According to paragraph .05 of AU-C section 530, *Audit Sampling* (AICPA, *Professional Standards*), audit sampling is "the selection and evaluation of less than 100 percent of the population of audit relevance such that the auditor expects the items selected (the sample) to be representative of the population and, thus, likely to provide a reasonable basis for conclusions about the population." It is often used to evaluate some characteristic of a balance or class of transactions and to obtain audit evidence. Auditors may use either nonstatistical or statistical sampling. The items selected for examination from the account balance or class of transactions is referred to as the *sample*. The entire set of data from which a sample is selected and about which the auditor wishes to draw conclusions is called the *population*.

How Audit Sampling Differs From Sampling in Other Professions

2.03 Auditing is not the only profession that uses sampling. For example, sampling is used in opinion surveys, market analyses, and scientific and medical research in which someone desires to reach a conclusion about a large body of data by examining only a portion of that data. There are major differences, though, between audit sampling as discussed in this guide and these other sampling applications.

2.04 Accounting populations differ from most other populations, because before the auditor's testing begins, the data have been accumulated, compiled, and summarized. Normally, the auditor's objective is to corroborate the accuracy of certain client data, such as data about account balances or classes of transactions, or to evaluate the effectiveness of controls in the processing of the data. The audit process is generally an evaluation of whether an amount is materially misstated rather than a determination of original amounts.

2.05 The distribution of amounts in some accounting populations may differ from other populations. In some nonaccounting populations, the amounts tend to cluster around the average amount of the items in the population. In contrast, many accounting populations tend to include a few very large amounts, a number of moderately large amounts, and a large number of small amounts. The auditor may need to consider the distribution of accounting amounts when planning audit samples for substantive procedures. For example, such information may be useful when stratifying the population or considering whether the audit sampling technique being used is likely to be effective in that population.

2.06 In addition, the evidence obtained from each audit test is just a portion of the total evidence that the auditor obtains. The auditor usually does not rely on a single audit test, as might a market researcher or another sampler, but reaches an overall conclusion based on the results of numerous interrelated tests that are performed. Therefore, an auditor plans and evaluates an audit sample with the knowledge that the overall conclusion about the population characteristic of interest is based on more than the results of that audit sample.

Evaluation of Audit Samples

2.07 AU-C section 530 and AU-C section 450, *Evaluation of Misstatements Identified in an Audit* (AICPA, *Professional Standards*), establish standards for audit sampling that apply to both statistical and nonstatistical sampling. These standards include the following:

- Where the item selected or the supporting documentation is not available to the auditor, the auditor usually treats the item as a deviation or misstatement. This presumption may be overcome by appropriate evidence.
- The auditor should project the results of the sample to the population from which the sample was selected, and not conclude solely on the specific sample deviations or factual misstatements (even if corrected by the client).
- The auditor should compare the projected deviation rate or misstatement to the tolerable rate or tolerable misstatement for the test of the account balance or class of transactions, and should appropriately consider sampling risk.
- The auditor should consider the qualitative aspects of the deviations or misstatements in assessing whether the evidence may suggest other issues that might alter the implied severity of the

assessment or need to be addressed in the audit. For example, a deviation might provide evidence of a fraud or a serious control issue.

Types of Audit Tests

2.08 AICPA *Professional Standards* describes several types of audit tests including tests of controls, substantive tests, and dual-purpose tests.[1] The type of test to be performed is important to an understanding of audit sampling.

Tests of Controls

2.09 Tests of controls provide evidence about the operating effectiveness of a control in preventing or detecting material misstatements in a financial statement assertion. In tests of controls, the auditor is usually concerned about the rates of any deviation from a prescribed control procedure. Tests of controls are necessary when the audit strategy is to rely on the effectiveness of the control. As discussed in the section "Some Tests of Controls May Not Involve Audit Sampling" in chapter 1, "Characteristics of Audit Sampling," some controls cannot be tested using audit sampling.

2.10 Controls typically are expected to be applied in the same way to all transactions subject to that policy or procedure, regardless of the magnitude of the transaction. Therefore, if the auditor is using audit sampling, it is usually not appropriate to select only high dollar amounts in tests of controls, unless the control is applied only to high dollar transactions. Sample items should be selected in such a way that the sample can be expected to be representative of the population, so that the auditor will be able to draw appropriate conclusions about the population.

Substantive Procedures

2.11 Substantive procedures are audit procedures designed to obtain evidence about the validity and propriety of the accounting treatment of transactions and balances or to detect misstatements.[2] Substantive procedures differ from tests of controls in that the auditor is interested primarily in a conclusion about dollars. Substantive procedures include (*a*) tests of details of transactions and balances and (*b*) analytical procedures.

Dual-Purpose Tests

2.12 In some circumstances, an auditor might design a test that has a *dual purpose*: testing the effectiveness of a control and testing whether a recorded balance or class of transactions is materially misstated. In using dual-purpose testing, an auditor may have begun substantive procedures before determining whether the test of controls supports the auditor's assessed level of control risk. Therefore, an auditor planning to use a dual-purpose sample will have made a preliminary judgment that there is an acceptably low risk that the rate of deviations from the prescribed control in the population exceeds the tolerable rate of deviations the auditor is willing to accept without altering the planned

---

[1] Dual-purpose tests are discussed in paragraph .A24 in the "Application and Explanatory Material" section of AU-C section 330, *Performing Audit Procedures in Response to Assessed Risks and Evaluating the Audit Evidence Obtained* (AICPA, *Professional Standards*). Applications of this concept often involve sampling applications.

[2] Substantive procedures may also reveal deficiencies in controls.

assessed level of control risk. For example, an auditor designing a test of the controls for entries in the voucher register might plan a related substantive procedures at a risk level that anticipates a particular assessed level of control risk. The assessed level of control risk would be dependent on the results of the test of the controls.

2.13 Assuming the same sample selection method is appropriate for both purposes, the size of a sample designed for a dual-purpose test will usually be the larger of the samples that would otherwise have been designed for the two separate purposes. In most cases, separate procedures (for example, tests of controls and substantive procedures) are applied to the common sample of transactions to draw both the control and substantive conclusions. The fact that a transaction was correctly processed substantively does not provide evidence that controls designed to achieve those objectives were in place and operating effectively. However, in some circumstances the performance of a single test may provide both substantive and controls evidence such as when reperforming a manual control that is designed to ensure clerical accuracy. The auditor ordinarily evaluates deviations from pertinent controls and monetary misstatements separately, using the risk level applicable for the respective purposes when evaluating dual-purpose samples. The guidance provided in chapters 3–7 for evaluating the results of tests of controls and substantive procedures is also applicable to the evaluation of dual-purpose samples.

2.14 When control and substantive sample sizes are very different due to the sampling parameters chosen, the auditor may consider whether the sample sizes can be made more similar by changing the audit strategy and balancing the reliance on controls versus the reliance on substantive procedures used in this situation. When the auditor believes that the use of the parameters resulting in very different sample sizes results in the best audit strategy, a dual-purpose test (common items identified for the two samples) can be accomplished by either testing both purposes with the larger sample or by first selecting the larger sample and then selecting an unbiased, representative selection of items from the larger sample to use for the smaller sample. For example, the smaller sample could be selected by taking a random, haphazard, or systematic (every nth item) sample from the larger sample. The subsample is usually not selected in such a way that the resultant sample can be expected to only represent a part of a year or be comprised of only very large items. This could happen, for example, if only the first items in a systematically selected larger sample or only the largest items are selected for the smaller subsample.

Risk

2.15 The justification for reasonable assurance (in other words, a high, but not absolute level of assurance) rather than certainty regarding the reliability of financial information is found in paragraph .05 of AU-C section 200, *Overall Objectives of the Independent Auditor and the Conduct of an Audit in Accordance With Generally Accepted Auditing Standards* (AICPA, *Professional Standards*), which states that

> as the basis for the auditor's opinion, GAAS require the auditor to obtain reasonable assurance about whether the financial statements as a whole are free from material misstatement, whether due to fraud or error. Reasonable assurance is a high, but not absolute, level of assurance.

2.16 The justification for accepting some uncertainty arises from the relationship between the cost and time required to examine all the data and the adverse consequences of possible erroneous decisions based on the conclusions resulting from examining only a sample of such data. The uncertainty inherent in performing auditing procedures is audit risk. At the account balance, class of transactions, relevant assertion, or disclosure level, audit risk consists of (*a*) the *risks of material misstatement* (consisting of inherent risk and control risk) and (*b*) detection risk.

2.17 According to paragraph .03 of AU-C section 315, *Understanding the Entity and its Environment and Assessing the Risks of Material Misstatement* (AICPA, *Professional Standards*),

> the objective of the auditor is to identify and assess the *risks of material misstatement*, whether due to fraud or error, at the financial statement and relevant assertion levels through understanding the entity and its environment, including the entity's internal control, thereby providing a basis for designing and implementing responses to the assessed *risks of material misstatement*.

2.18 It is not acceptable to simply deem risk to be "at the maximum." This assessment may be in qualitative terms such as high, medium, and low, or in quantitative terms such as percentages. Audit risk includes uncertainties due to both sampling and other factors. These are sampling risk and nonsampling risk, respectively.

Sampling Risk

2.19 Sampling risk is the risk that the auditor's conclusion based on a sample might be different from the conclusion he or she would reach if the test were applied in the same way to the entire population. Sampling risk arises from the possibility that a particular sample might contain proportionately more or less monetary misstatement or deviation from prescribed controls than exist in the account balance or class of transactions as a whole. Sampling risk includes the risk of concluding that controls are more effective than they actually are and the risk of concluding that controls are less effective than they actually are (see discussions in chapter 1 and chapter 3, "Nonstatistical and Statistical Audit Sampling in Tests of Controls") as well as the risk of incorrect acceptance and the risk of incorrect rejection for substantive procedures (see discussions in chapter 1 and chapter 4, "Nonstatistical and Statistical Audit Sampling for Substantive Tests of Details").

Nonsampling Risk

2.20 Paragraph .05 of AU-C section 530 defines nonsampling risk as the risk that the auditor reaches an erroneous conclusion for any reason not related to sampling risk." Nonsampling risk includes all the aspects of audit risk that are not due to sampling. An auditor might apply a procedure to all transactions or balances and still fail to detect a material misstatement or the ineffectiveness of a control. Nonsampling risk includes the possibility of using audit procedures that are not appropriate to achieve the specific objective. For example, the auditor cannot rely on confirmation of recorded receivables to reveal whether there are unrecorded receivables. Nonsampling risk also arises because the auditor might fail to recognize deviations or misstatements included in documents that he or she examines. In that situation, the audit procedure would be ineffective even if all items in the population were examined.

2.21 There is no common method that allows the auditor to measure nonsampling risk. This risk can, however, be reduced to an acceptable level by adequate planning and supervision of audit work (see AU-C section 300, *Planning an Audit* [AICPA, *Professional Standards*]) and by implementing an effective quality control system (see AU-C section 220, *Quality Control for an Engagement Conducted in Accordance With Generally Accepted Auditing Standards* [AICPA, *Professional Standards*]). Also, the auditor ordinarily considers nonsampling risk when designing his or her audit procedures. If there is a *choice* of audit procedures, both of which provide the same level of assurance at approximately the same cost, the auditor ordinarily uses the procedure with the lower nonsampling risk. The subject of controlling nonsampling risk is beyond the scope of this guide; however, the "General Implementation Considerations" section of this chapter might be helpful to the auditor in controlling some aspects of nonsampling risk.

Nonstatistical and Statistical Sampling

2.22 All audit sampling involves judgment in planning and performing the sampling procedure and evaluating the results of the sample. The audit procedures performed in examining the selected items in a sample typically do not depend on the sampling approach used.

2.23 Once a decision has been made to use audit sampling, the auditor may choose to use either statistical or nonstatistical sampling. This choice is often a cost-benefit consideration. Statistical sampling helps the auditor (*a*) design an efficient sample, (*b*) measure the sufficiency of the audit evidence obtained, and (*c*) quantitatively evaluate the sample results. If audit sampling is used, some sampling risk is always present. Statistical sampling uses the laws of probability to measure sampling risk. Any sampling procedure that does not permit the numerical measurement of the sampling risk is a nonstatistical sampling procedure. Even though the auditor rigorously selects a random sample, the sampling procedure is a nonstatistical application if the auditor does not make a statistical evaluation of the sample results.

2.24 A properly designed nonstatistical sampling application that considers the same factors that would be considered in a properly designed statistical sample can provide results that are as effective as those from a properly designed statistical sampling application; however, there is one important difference: statistical sampling explicitly measures the sampling risk associated with the sampling procedure by providing an explicit level of sampling risk (also sometimes expressed as its complement—confidence or reliability) and allowance for sampling risk (that is, precision) about the sample result.

2.25 Statistical sampling might involve different training because it requires more specialized expertise. The use of audit sampling software can reduce the costs of applying statistical sampling. Such software is commonly used to select random, systematic, or stratified samples whether or not the sample is statistically evaluated.

2.26 However, it may not be efficient to use sampling software when the population is not already in electronic format. For example, if the individual balances constituting an account balance to be tested are manual records and not maintained in an organized pattern, it might not be efficient for an auditor to select items in a way that would satisfy the requirements of a properly designed statistical sample. In such a circumstance, that auditor will still need to obtain evidence that the population is complete and that determination may provide a suggested approach for sample selection.

2.27 Another example of when it may be difficult to apply statistical sampling is when the auditor plans to use audit sampling to test a physical inventory count and the client does not maintain perpetual inventory records. Although the auditor can select a sample so that the sample can be expected to be representative of the population (selected without bias), it might be difficult to satisfy certain requirements for a statistical sample if priced inventory listings or detailed prenumbered quantity listings cannot be used in the selection process. (See the section "Determining the Method of Selecting the Sample" in chapter 3.) Because either nonstatistical or statistical sampling can provide sufficient audit evidence, the auditor chooses between them after considering their relative efficiency and effectiveness in the circumstances.

2.28 Statistical sampling provides the auditor with a tool that assists in applying experience and professional judgment to explicitly control sampling risk. Because this risk is present in both nonstatistical and statistical sampling plans, there is no conceptual reason to expect a nonstatistical sample to provide different assurance from a well-designed statistical sample of comparable size for the same sampling procedure.[3] AU-C section 530 states that the sample size of a nonstatistical sample would ordinarily be comparable to the sample size resulting from an efficient and effectively designed statistical sample, (considering the same sampling parameters); however, neither AU-C section 530 nor this guide requires the auditor using nonstatistical sampling to compute a sample size using statistical theory when determining the sample size for the nonstatistical sampling application.

2.29 With nonstatistical sampling the auditor normally relies on professional judgment, in combination with nonstatistical sampling guidance and knowledge underlying statistical concepts, to design and evaluate audit samples. A risk associated with nonstatistical sampling is that the auditor's judgment may diverge significantly from sampling concepts resulting in testing that is not as effective as statistical sampling.[4] Some auditors address this risk by providing audit staff with nonstatistical sampling guidance and procedures that are easy to use, encourage consistency in sampling applications across engagement teams, and are grounded in sampling theory.

Planning the Audit Sampling Procedures

2.30 When an auditor plans any audit sampling application, the first consideration is the specific account balance or class of transactions and the circumstances in which the procedure is to be applied. The auditor normally identifies items or groups of items that are of individual significance to an audit objective or assertion. For example, an auditor planning to use audit sampling as part of a substantive procedure for an inventory balance, including observing

---

[3] Chapters 3–7 provide several quantitative illustrations of sample sizes based on statistical theory. They may be helpful to an auditor applying professional judgment and experience in considering the effect of various planning considerations on sample size.

[4] There is also a potential risk that auditors may misapply statistical concepts.

the physical inventory, would normally identify items that have significantly large balances or that might have other special characteristics (such as higher susceptibility to obsolescence or damage). In testing accounts receivable, an auditor might identify accounts with large balances, unusual balances, higher risks, or unusual patterns of activity as individually significant items.

2.31 A best practice is for the auditor to consider all special knowledge about the items constituting the balance or class before designing audit sampling procedures. For example, the auditor might identify 20 products included in the inventory that make up 25 percent of the account balance. In addition, he or she might have identified several items, constituting an additional 10 percent of the balance that are especially susceptible to damage. The auditor might decide that those items, comprising 35 percent of the balance should be examined 100 percent and therefore need not be included in the inventory subject to audit sampling.

2.32 After the auditor has applied any special knowledge about the account balance or class of transactions in designing an appropriate procedure, often a group of items remains that needs to be evaluated to achieve the audit objective. Thus in the preceding example, the auditor might apply audit sampling, either nonstatistical or statistical, to the remaining 65 percent of the account balance. The considerations just described would not be influenced by the auditor's intentions to use either nonstatistical or statistical sampling on the remaining items.

2.33 The following questions apply to planning any audit sampling procedure, whether it is nonstatistical or statistical:

- What is the test objective and relevant assertion? (What does the auditor want to learn or be able to infer about the population? What assertions are being tested?)
- What is the auditor looking for in the sample? (How is a misstatement or deviation defined?)
- What is to be sampled and is the population complete? (How is the population defined?). Is the population from which the sample is selected the same population the sample results will be projected to?[5]
- How is the population to be sampled? (What is the sampling plan, what is the sampling unit, and what is the method of selection?)
- How much is to be sampled? (What is the sample size?)
- What do the results mean? (How are the sample results evaluated and interpreted?)

2.34 As discussed in chapter 1, audit sampling may not always be efficient or appropriate. For example, the auditor might decide that it is more efficient to test an account balance or class of transactions by applying only analytical procedures (assuming the assertions in the account have not been identified as a significant risk, analytical procedures should be supplemented with other procedures, such as substantive tests of details, control tests, or both). In some cases, legal or regulatory requirements might necessitate 100 percent examination. In other situations, the auditor might decide that some items should

---

[5] It is observed in practice that it may be important that care be taken that the population from which the sample is selected is the same population used for the projection from the sample result.

be examined 100 percent because he or she does not believe acceptance of sampling risk is justified, or he or she believes a 100 percent examination is more efficient in the circumstances. The auditor uses professional judgment to determine whether audit sampling is appropriate.

Types of Statistical Sampling Plans

Attributes Sampling

2.35 Attributes sampling is used to reach a conclusion about a population in terms of a rate of occurrence. Its most common use in auditing is to test the rate of deviation from a prescribed control to support the auditor's assessed level of control risk. In attributes sampling,[6] each occurrence of, or deviation from a prescribed control, is given equal weight in the sample evaluation, regardless of the dollar amount of the transactions. For testing the operating effectiveness of controls that are expected to operate with the same level of consistency, regardless of the size of transactions, attributes sampling is typically the most effective method for applying audit sampling to these tests.

2.36 Some examples of tests of controls in which attributes sampling is typically used include test of controls over the following:

- Voucher processing
- Billing systems
- Payroll and related personnel-policy systems

In most cases, manual control activities are generally susceptible to attributes sampling.

2.37 In addition to tests of controls, attributes sampling may be used as substantive procedures, such as tests for under-recorded shipments or understated demand deposit accounts, when the objective is to determine whether proper revenue recognition or cut-off occurred, and no misstatements or deviations are anticipated; however, if the audit objective is to obtain evidence directly about a monetary amount being examined, such that the sample result may be projected in monetary terms, the auditor usually designs a variables sampling application.

Variables Sampling

2.38 Variables sampling is used if the auditor desires to reach a conclusion about a population in terms of a dollar amount. Variables sampling is typically used to answer either of these questions:

  a. How much? (sometimes described as dollar-value estimation)
  b. Is the account materially misstated? (sometimes described as hypothesis testing).

Both monetary unit sampling (MUS), discussed in chapter 6, "Monetary Unit Sampling," and classical variables sampling, discussed in chapter 7, "Classical Variables Sampling," are examples of variables sampling.

2.39 The principal use of variables sampling in auditing is to substantively test details to determine the reasonableness of recorded amounts; however, it

---

[6] As used in this guide, attributes sampling refers to unstratified attributes sampling. Stratified attributes sampling is not discussed in this guide.

might also be used if the auditor chooses to estimate the dollar amount of transactions containing deviations from a control (see footnote 2 of chapter 6), such as when assessing the severity of a deficiency in controls.

2.40 Some examples of tests for which variables sampling is typically used include tests of the following:

- The existence of valid receivables
- The accuracy of inventory quantities and amounts
- The occurrence of recorded payroll expense
- The existence of fixed-asset additions

2.41 Attributes sampling is frequently used to reach a conclusion about a population in terms of a rate of occurrence; variables sampling is frequently used to reach conclusions about a population in terms of a dollar amount. MUS is based on attributes sampling theory, but is applied as a variables sample and is able to express conclusions in monetary terms.

Relating Balance Sheet and Income Statement Sampling

2.42 Accounts in the balance sheet and income statement are often related. Auditors, in obtaining direct assurance with respect to certain balance sheet accounts (for example, through confirmations of accounts receivables and performance of cash reconciliations), often also obtain some assurance through such testing on some assertions in the related income statement accounts.[7] For example, auditors who obtain direct assurance from tests regarding the existence of accounts receivable and completeness and occurrence of cash collections, often also obtain some assurance from these balance sheet tests regarding the occurrence assertion in the revenue accounts. The nature and extent of the tests performed on related balance sheet accounts (for example, receivables), in addition to any other evidence obtained regarding the relevant assertions in related income statement accounts, such as the revenues account, may be considered when determining whether additional audit evidence regarding one or more assertions needs to be obtained from direct tests of income statement accounts such as revenues.

2.43 In some cases, the audit procedures performed on balance sheet accounts may not sufficiently address the relevant assertions and risks in related income statement accounts. For example, suppose an identified revenue risk was that the *custom* contractual terms in machine and maintenance sales agreements could require a different generally accepted accounting principles (GAAP) treatment (for example, a portion of the revenue should be deferred) that might not be reflected properly in the accounting records. If the procedures performed on the receivables and cash receipts did not adequately address this risk, then additional tests involving a sample of revenue transactions may be needed to reduce the risk of a material misstatement in revenues related to realization to *low*. In other situations, all revenue transactions may have similar contractual terms that result in a clear and consistent GAAP treatment, and such risk might not be present. When determining the nature, timing, and

---

[7] Similarly, direct tests of the income statement accounts often provide some evidence regarding the related balance sheet accounts. Readers may also find further discussion of the use of assertions in auditing both balance and transaction data in paragraphs 2.28–.34 and table 2-3 of the AICPA Audit Guide *Assessing and Responding to Audit Risk in a Financial Statement Audit*.

extent of procedures performed on the income statement accounts, the auditor would normally consider the risks and evidence obtained or planned to be obtained from other audit procedures related to the assertions relevant to the income statement account.

General Implementation Considerations

2.44 Consideration of the following factors might be helpful in implementing audit sampling procedures.

Continuing Professional Education

2.45 Audit sampling and the concepts of statistical sampling are topics that have appeared in the CPA examination for decades. Many college auditing courses and auditing textbooks cover the principles of sampling as applied in auditing. Many business degree programs also require a course on the application of probability and statistics to business data.

2.46 The auditor may better understand the application of the concepts of audit sampling by combining live instruction with this guide or a textbook. Some auditors attend continuing professional educational (CPE) programs developed by their firms, whereas others attend such programs developed by the AICPA, a state society of CPAs, a college or university, a software vendor or another CPA firm.

2.47 Relevant CPE programs are normally directed to appropriate professional personnel. For example, a firm might decide to train all audit personnel to select samples, determine sample sizes, and evaluate sample results for attributes sampling procedures. More experienced audit personnel might be trained to design and evaluate variables sampling applications.

2.48 Because of the computational aspects of statistical sampling and the availability of computer programs to design and perform a sample, courses in applying statistical sampling often include training in the use of software and practice aids and focus on using software or tables for determining sample size, selecting the sample, and drawing a statistical conclusion from the sample results.

Sampling Guidelines

2.49 Some auditors achieve greater consistency in sampling applications throughout their practices by establishing sampling guidelines, such as guidelines about acceptable risk levels, minimum sample sizes, and appropriate levels of tolerable misstatement.

Use of Specialists

2.50 Because statistical sampling concepts are well established as a subject area of desired competence for certification as a CPA, auditors ordinarily will have the ability to apply basic statistical concepts and procedures to audit situations when the occasion arises. Some auditors designate selected individuals within their firm as audit sampling specialists.[8] These specialists may

---

[8] An audit sampling specialist who is a member of the audit staff is considered part of the engagement team. AU-C section 620, *Using the Work of an Auditor's Specialist* (AICPA, *Professional Standards*), establishes requirements and provides guidance when the auditor uses the work of a specialist.

consult with other audit personnel on the design and execution of planned sampling procedures. In addition, some specialists teach CPE courses on audit sampling. Some firms train all audit personnel in the essential concepts of designing and executing sampling procedures, thus minimizing the need for specialist assistance on most engagements.

2.51 Furthermore, some auditors also engage an outside consultant for certain statistical applications. The consultant might (*a*) assist in solving difficult statistical problems arising in practice, (*b*) review sampling guidelines and methodologies, (*c*) assist in designing CPE programs, and (*d*) teach courses for specialists.

Supervision and Review

2.52 Paragraph .11 of AU-C section 300 states that assistants should be properly supervised. When establishing the overall strategy for the audit, the auditor determines a materiality level for the financial statements as a whole and may quantify measurements of risk. Use of quantifiable concepts, even though subjective, can be useful in communicating audit objectives to the auditor's assistants.

2.53 Review of documentation of audit sampling procedures designed by assistants in the planning stage helps to determine that the application has been well planned and can be implemented successfully. Review of the work and evaluation provides comfort that the work has been done properly and the conclusions are appropriate.

2.54 In reviewing audit sampling applications, the auditor might consider the following questions:

- Was the test objective appropriate?
- Were the population and sampling unit (and relevant assertion) defined appropriately for the test objective?
- Were misstatements or deviations defined appropriately?
- Were tests performed to provide reasonable assurance that the sample was selected from the appropriate population?
- Did the design of the sampling application provide for an appropriate risk level? For example, did the design reflect the auditor's assessed level of the *risks of material misstatement* and the desired evidence to be obtained from related substantive procedures?
- If additional substantive procedures (for example, analytical procedures) were planned in designing the sampling procedure, did these tests support the assertions about the account being tested?
- Were planned procedures applied to all sample items? If not, were unexamined items considered in the evaluation?
- Were all deviations or misstatements discovered properly evaluated? For example were missing items properly evaluated, were the misstatements projected and evaluated properly along with the associated sampling risk, and was the nature of the misstatements properly considered?
- If the test was a test of controls, did it support the planned assessed level of control risk? If not, were related substantive procedures appropriately modified?

AAG-SAM 2.51

- If the test was a substantive procedure, did it support the relevant assertion(s) for the account balance or class of transactions? If not, were appropriate steps taken?
- Was the audit objective of the test met?

2.55 The general concepts discussed in this chapter are applied to tests of controls and substantive procedures in chapters 3–4, respectively.

Chapter 3

Nonstatistical and Statistical Audit Sampling in Tests of Controls

> **⊙ Update 3-1 *Audit*: Clarified Auditing Standards**
>
> The auditing guidance in this guide edition has been conformed to Statement on Auditing Standards (SAS) Nos. 122–125, which were issued in 2011 as part of the Auditing Standards Board's Clarity Project. These clarified SASs are effective for audits of financial statements for periods ending on or after December 15, 2012. Although extensive, the revisions to generally accepted auditing standards resulting from these clarified SASs do not change many of the requirements found in the auditing standards that they supersede.
>
> To assist auditors and financial reporting professionals in making the transition, this guide includes appendix F, "Mapping and Summarization of Changes—Clarified Auditing Standards," which provides a cross reference of the sections in the superseded auditing standards to the applicable sections in the clarified auditing standards and identifies the changes, either substantive or primarily clarifying in nature, that may affect an auditor's practice or methodology relative to the applicable sections of SAS Nos. 122–125. It also summarizes the changes resulting from the requirements of SAS Nos. 122–125.
>
> The preface of this guide and the Financial Reporting Center on www.aicpa.org provide more information on the Clarity Project. Visit www.aicpa.org/sasclarity.

3.01 This chapter introduces the general concepts of audit sampling applicable to statistical and nonstatistical sampling for tests of controls. It also discusses guidelines for determining the sample size and performing the sampling plan and evaluating the results of applying audit procedures.

Determining the Test Objectives

3.02 As mentioned in chapter 2, "The Audit Sampling Process," the objective of tests of controls is to provide evidence about the operating effectiveness of controls. The auditor performs tests of controls to support his or her assessed level of control risk. Tests of controls, therefore, are concerned primarily with these questions:

    a. Were the necessary controls performed?

    b. How were they performed?

    c. By whom were they performed?

3.03 AU-C section 315, *Understanding the Entity and Its Environment and Assessing the Risks of Material Misstatement*; AU-C section 330, *Performing Audit Procedures in Response to Assessed Risks and Evaluating the Audit Evidence Obtained* (AICPA, *Professional Standards*); and the AICPA Audit Guide *Assessing and Responding to Audit Risk in a Financial Statement Audit* provide guidance on identifying relevant controls and designing and evaluating the results of tests of controls.

3.04 Audit sampling for tests of controls is generally appropriate when application of the control leaves documentary evidence of performance. Audit sampling for tests of controls that do not leave such evidence (such as some automated controls) might be appropriate, however, when the auditor is able to plan the audit sampling procedures early in the engagement. For example, the auditor might wish to observe the performance of prescribed control activities for bridge toll collections. In that case, a sample of days and locations for observation of actual activities would be selected. The auditor needs to plan the sampling procedure to allow for observation of the performance of such activities on days selected from the period under audit.

3.05 When the auditor seeks an understanding of internal controls, evidence that the control has been implemented (placed in operation) is generally obtained by observing, performing walkthroughs, or examining one or a few instances of the control's operation. The auditor documents the evidence obtained supporting his or her conclusions that the controls are in place. Applying audit sampling may not be necessary when selecting just one or a few items for inspection if the purpose is to obtain evidence about those items rather than to reach a conclusion about the population.

Defining the Deviation Conditions

3.06 Based on the auditor's understanding of internal control, he or she will generally identify the characteristics that would indicate performance of the control to be tested. The auditor then defines the possible deviation conditions. For tests of controls, a deviation is a departure from the expected performance of the prescribed control. Performance of a control consists of all the steps the auditor believes are necessary to support his or her assessed level of control risk. For example, a prescribed control may require that disbursements are supported by an invoice, a voucher, a receiving report, and a purchase order, all stamped *Paid*. In this case, a deviation may be defined as "a disbursement not thus supported." Once the auditor has established that the *Paid* stamp does in fact indicate that the control has been performed (for example, testing a few instances that the presence of the stamp properly indicates the operation of the control), the operating effectiveness of the control may be further tested by sampling disbursements and noting the presence or absence of the *Paid* stamp.

Defining the Population

3.07 The population consists of the items constituting the account balance or class of transactions of interest. The auditor should determine that the population from which the sample is selected is appropriate for the specific audit objective, because sample results can be projected only to the population from which the sample was selected. For example, if the auditor wishes to test the operating effectiveness of a prescribed control designed to ensure that all shipments are billed, it would be ineffective to sample items that have already been billed. Rather, the auditor generally would sample the population of shipped items to determine whether selected shipments were billed.

3.08 An auditor is generally alert to the possibility that an entity might change a specific control during the period under audit. If one control is superseded by another that is designed to achieve the same control objective, the auditor needs to decide whether to test the operating effectiveness of both controls, or only the more recent one. This depends on the auditor's objective. For example, if the auditor requires evidence about the operating effectiveness of both the new and the old control to support an assessed level of control risk and the old and new procedures are both expected to be effective, a sample of all sales transactions may be appropriate. Auditors might also design two separate samples to accomplish the audit objective, especially where the controls are significantly different. However, if the auditor's assessment of control risk is primarily dependent on effective application of controls in the latter part of the period or as of a specific point in time, he or she might obtain evidence about the operating effectiveness of the new control mainly or exclusively, and obtain little or no evidence about the superseded control. In designing an appropriate sample, the auditor considers what is effective and efficient in the circumstances. For example, if the auditor wishes to test both old and new controls, it may be more efficient, yet still effective, to design one sample of all such transactions executed throughout the period than to design separate tests of the transactions subject to the two different controls.

3.09 For example, if the auditor desires to conclude on the effectiveness of controls during a reporting period in order to rely on those controls for the financial statement audit and a new computer system over revenue was installed mid-year, it would be necessary to test controls from both systems in order to obtain evidence about the controls' effectiveness over the entire period; however, if a new system is installed to replace one demonstrated or known to be ineffective, reliance on the ineffective system during its period of operation is not warranted.

3.10 If an attest engagement (for example, AT section 501, *An Examination of an Entity's Internal Control Over Financial Reporting That Is Integrated With an Audit of Its Financial Statements* [AICPA, *Professional Standards*]) to report on the effectiveness of controls is expressed "as of" a specific date, tests of controls are designed to principally relate to controls in effect as of the reporting date.

Defining the Period Covered by the Test

3.11 When an auditor performs tests of controls during interim work, he or she should consider what additional evidence needs to be obtained for the remaining period. Paragraph .12 of AU-C section 330 establishes requirements for the auditor when the operating effectiveness of controls has been tested during an interim period. Where this is obtained by extending the test to transactions occurring in the remaining period, the population consists of all transactions executed throughout the period under audit. If the test is not extended, the population consists only of transactions for the interim period and the results of the test can only be projected to that period. In this case, the auditor obtains other evidence to conclude on the operating effectiveness of those controls during the period not covered by the tests of controls. In determining the nature and extent of these additional tests, the auditor considers the following factors, which are enumerated in paragraph .A36 of AU-C section 330, in determining what, if any, additional evidence needs to be obtained for the remaining period:

- The significance of the assessed risks of material misstatement at the relevant assertion level

- The specific controls that were tested during the interim period and the results of those tests
- Significant changes to the controls since they were tested, including changes in the information system, processes, and personnel
- The degree to which audit evidence about the operating effectiveness of those controls was obtained
- The length of the remaining period
- The extent to which the auditor intends to reduce further substantive procedures based on the reliance of controls
- The effectiveness of the control environment

3.12 The auditor obtains evidence about the nature and extent of any significant changes in internal control, including personnel performing the control, which occur during the remaining period. If significant changes do occur, the auditor considers the effects on the audit strategy and audit plan, and may revise his or her understanding of internal control and consider testing the changed controls. Alternatively, the auditor may consider performing substantive analytical procedures or tests of details covering the remaining period.

3.13 When the auditor requires assurance regarding the effectiveness of controls as of a specific date (for example, an attestation engagement to report on the effectiveness of internal controls, described in AT section 501), the transactions on or close to that date constitute the population from which a sample is selected. When it is impractical to perform tests on controls in that period, it may be appropriate to test controls in operation at an earlier period provided that (*a*) effective IT general controls exist and are tested to support reliance on the proper operation of the control throughout the period, (*b*) there is evidence that the control procedure has not changed, and (*c*) the auditor updates the understanding and testing results to the "as of" date. Procedures to update controls assessments through the year include inquiry, combined with corroborating evidence provided by observation, walkthroughs, or additional control tests performed close to the "as of" date.

Initial Testing

3.14 The auditor might define the population to include transactions from the entire period under audit, but perform initial testing during an interim period. In such circumstances, the auditor would often estimate the number of transactions that will be executed during the remaining period and design the sample based on that estimate. For example, if in the first 10 months of the year, the entity issued invoices numbered from 1 to 10,000, the auditor might estimate that another 2,500 invoices will be issued in the last 2 months and use 1 to 12,500 as the numerical sequence for selecting the desired sample. Invoices with numbers 1 to 10,000 would be subjected to possible selection during the interim work, and the remaining 2,500 invoices would be subject to sampling during the completion of the audit.

Estimating Population Characteristics

3.15 In estimating the size of the population, the auditor might consider such factors as the actual usage in the similar period of the prior year, the trend of usage, and the nature of the business. As a practical consideration, the auditor might overestimate the remaining volume. If at year-end some of the selected document numbers do not represent executed transactions (because fewer transactions were executed than estimated), they may be replaced by other transactions. To provide for this possibility, the auditor might select a slightly larger number of items than indicated by the minimum sample size; the additional items would be examined only if they are needed as replacement items.

3.16 If, on the other hand, the remaining usage is underestimated, some transactions will not have a chance of being selected and the sample would not have been selected from the population defined by the auditor. In this case, the auditor may redefine the population to formally exclude those items not included in the population for sampling. In the latter case, the auditor may then perform alternative procedures to reach a conclusion about the items not included in the redefined population. Such tests might include testing the items as part of a separate sample, examining 100 percent of the items, or making inquiries and observations as well as obtaining some additional evidence concerning the remaining period. The auditor determines an appropriate approach based on his or her judgment about which procedure would be effective and efficient in the circumstances.

3.17 In some cases, the auditor might not need to wait until the end of the period under audit to form a conclusion about whether the operating effectiveness of a control supports his or her planned assessed level of control risk. During the interim testing of selected transactions, the auditor might discover deviations sufficient to reach the conclusion that, even if no deviations are found in transactions to be executed after the interim period, the control would not support the planned assessed level of control risk. In that case, the auditor might decide not to extend the sample to transactions to be executed after the interim period and would modify the nature, timing and extent of planned substantive procedures accordingly. Significant deficiencies and material weaknesses must be reported to management and those charged with governance in writing, as described in paragraph .11 of AU-C section 265, *Communicating Internal Control Related Matters Identified in an Audit* (AICPA, *Professional Standards*).

Considering the Completeness of the Population

3.18 The auditor selects sampling units[1] from a physical representation of the population. For example, if the auditor defines the population as all customer receivable balances as of a specific date, the physical representation might be the printout of the customer accounts-receivable trial balance as of that date or an electronic file purportedly containing the customer balances. Alternatively, the population may be defined as all unpaid invoices as of a specific date.

3.19 The auditor should consider whether the physical representation includes the entire population. Because the auditor actually selects a sample from the physical representation, any conclusions based on the sample relate only to that physical representation. If the physical representation and the desired population differ, the auditor might make erroneous conclusions about the population. For example, if the auditor wishes to perform a test of controls for the vouchers issued in 20XX, such vouchers are the population. If the auditor

---

[1] A sampling unit is any of the individual items (elements) constituting the population.

physically selects the vouchers from a filing cabinet, the vouchers in the filing cabinet are the physical representation. If the vouchers in the cabinet represent all the vouchers issued in 20XX, the physical representation and the population are the same. If they are not the same because vouchers have been removed or vouchers issued in other years have been added, the conclusion applies only to the vouchers in the cabinet.

3.20 Making selections from a controlled source minimizes differences between the physical representation and the population. For example, an auditor sampling vouchers might make selections from a voucher register or a cash disbursements journal that has been reconciled with issued checks by a comparison with open vouchers or through a bank reconciliation. The auditor might test the footing to obtain reasonable assurance that the source of selection contains the same transactions as the population.

3.21 If the auditor determines that items are missing from the physical representation, then the auditor would select a new physical representation or perform alternative procedures on the missing items. As a best practice, the auditor also would usually inquire about the reason that items are missing.

Defining the Sampling Unit

3.22 A sampling unit for tests of controls may be, for example, a document, an entry, or a line item where examination of the sampling unit provides evidence of the operation of the control. Each sampling unit constitutes one item in the population. The auditor may define the sampling unit in light of the control being tested. For example, if the test objective is to determine whether disbursements have been authorized and the prescribed control requires an authorized signature on the voucher before processing, the sampling unit might be defined as the voucher. On the other hand, if one voucher pays several invoices and the prescribed control requires each invoice to be authorized individually, the line item on the voucher representing the invoice might be defined as the sampling unit. Note that each sampling unit may provide evidence of the application of more than one control. For example, support for recording a receivable may indicate that the billed service was rendered or product shipped, the amounts were checked for accuracy, and the customer is listed on the approved customer list.

3.23 An overly broad definition of the sampling unit might not be efficient. For example, if the auditor is testing a control over the pricing of invoices and each invoice contains up to ten items, the auditor could define the sampling unit as an individual invoice or as a line item on the invoice. If the auditor defines the invoice as the sampling unit, the auditor would test all the line items on the invoice. If the auditor defines the line items as the sampling unit, only the selected line items need tested. If either sampling unit definition is appropriate to achieve the test objective, it is commonly more efficient to define the sampling unit as the more detailed alternative, in this case, a line item.

3.24 An important efficiency consideration in selecting a sampling unit is the manner in which the documents are filed and cross-referenced. For example, if a test of purchases starts from the purchase order, it might not be possible to locate the voucher and canceled check in some accounting systems because the systems have been designed to provide an audit trail from voucher to purchase order, but not necessarily vice versa.

The Role of Walkthroughs

3.25 A walkthrough of a transaction process does not involve audit sampling, as discussed in chapter 1, "Characteristics of Audit Sampling." A walkthrough is generally designed to provide evidence regarding the design and implementation of controls.[2] However, a walkthrough may be designed to include procedures that are also tests of the operating effectiveness of relevant controls (for instance, inquiry combined with observation, inspection of documents, or reperformance). If such procedures are performed in the context of a walkthrough, the auditor usually considers whether the procedures are performed at an adequate level to obtain some evidence regarding the operating effectiveness of the control. Such a determination would depend on the nature of the control (for example, automated versus manual), and on the nature of the auditor's procedures to test the control (for example, inquiry about the entire year and observation versus examination of documents or reperformance). For example, when a walkthrough includes inquiry and observation of the people involved in executing a control and where the auditor is satisfied that a strong control environment and adequate monitoring are in place, the auditor may conclude that the process provides some evidence about operating effectiveness. The auditor uses professional judgment to evaluate the extent of evidence obtained. In some cases, the procedures performed during the walkthrough may provide sufficient evidence of operating effectiveness (for example, for a fully automated control procedure in a system with effective IT general controls). In other cases, the auditor may conclude that the procedures performed during the walkthrough provide evidence to reduce but not eliminate other control testing; in those situations, the auditor might consider using a higher risk of overreliance (a lower confidence level) in designing these other control tests. The auditor needs to consider the evidence obtained from the design assessment and walkthrough and may use that information when determining the additional testing or procedures necessary to conclude on the sufficiency of audit evidence relative to the operating effectiveness of the controls.

3.26 When the auditor has performed only an assessment of design and implementation and assessed the design as effective and has obtained evidence that the controls have been implemented, the auditor might use a slightly lower confidence level for detailed substantive procedures (for example, 92 percent or 93 percent rather than a 95 percent confidence level if that was the level that the auditor would have otherwise planned for tests of details had the design or implementation of controls been assessed as ineffective).

3.27 If the auditor performs procedures that are a test of operating effectiveness of a control as part of a walkthrough, the auditor usually considers the extent of additional instances of the operation of the control that need to be examined to allow a conclusion regarding the control's operating effectiveness at the level of desired reliance. For automated controls, the walkthrough may sometimes be sufficient evidence when the IT general controls are effective.

3.28 If an audit sample of repeated occurrences of a control is deemed necessary (for example, examining documentation relating to a manual control), the test of controls performed in the context of the walkthrough is generally considered to yield the assurance regarding operating effectiveness that comes

---

[2] In prior AICPA literature the term *implementation* was stated as "placed in operation."

from a sample size of one for each item walked through the system. In such circumstances, the auditor may select an audit sample to gather evidence relating to additional instances of the operation of the control in order to obtain a significant level of assurance relating to operating effectiveness. When repeated instances of a control's execution are required to draw a conclusion regarding operating effectiveness, the evidence obtained in the context of the walkthrough is generally insufficient to conclude that the control is operating effectively.

Determining the Method of Selecting the Sample

3.29 Sample items should be selected so the sample can be expected to be representative of the population and thus the results can be projected to the population. Therefore, all items in the population should have an opportunity to be selected. These principles apply whether one applies nonstatistical or statistical sampling. For statistical sampling, it is necessary to use an appropriate random sampling method such as simple random sampling or systematic random sampling. In nonstatistical sampling, the auditor uses a sample selection approach that approximates a random sampling approach (for example, haphazard selection). Computer assisted audit technique (CAAT) software, as well as more general purpose spreadsheet software may be used to efficiently select statistical samples. An overview of selection methods follows.

Simple Random Sampling

3.30 With this method, every combination of sampling units has the same probability of being selected as every other combination of the same number of sampling units. To perform this selection, the auditor may select a random sample by matching random numbers generated by a computer or selected from a random-number table with, for example, document numbers. This approach is appropriate for both nonstatistical and statistical sampling applications.

Systematic Sampling

3.31 For this method, the auditor determines a uniform interval by dividing the number of physical units in the population by the sample size. A starting point is randomly selected in the first interval and one item is selected throughout the population at each of the uniform intervals from the starting point. For example, if the auditor wishes to select 100 items from a population of 20,000 items, the uniform interval is every 200th item. The auditor selects the first item from within the first interval and then selects every 200th item from the starting point. When the first item is selected randomly from the interval, the technique is called systematic random sampling. Paragraph .A16 of AU-C section 530, *Audit Sampling* (AICPA, *Professional Standards*), provides guidance on random selection techniques.

3.32 When a random starting point is used, the systematic method provides a sample that allows every sampling unit in the population an equal chance of being selected. If the population is arranged randomly with respect to its deviation pattern, systematic selection is equivalent to simple random selection. In the absence of a known pattern in the population, it is a practical and efficient alternative to simple random selection, particularly when items are being selected manually from a population. A potential problem with

Nonstatistical and Statistical Audit Sampling in Tests of Controls

systematic sampling is that the selection interval may coincide with a pattern in the population, thus biasing the selection. For example, a population of employees on a payroll for a construction company might be organized by teams; each team consists of a crew leader and nine other workers. A selection of every tenth employee on a sequential list of payroll payments will either list every crew leader or no crew leaders, depending on the random start point. No combination would include both crew leaders and other employees. In these circumstances, the auditor may consider using a different sample selection method, such as simple random number selection, or making a systematic selection using two or more random starting points or using an interval that does not coincide with a known pattern in the population.[3]

Haphazard Sampling

3.33 A *haphazard sample* is a nonstatistical sample selection method that attempts to approximate a random selection by selecting sampling units without any conscious bias, that is, without any special reason for including or omitting items from the sample. It does not imply the sampling units are selected in a careless manner; rather, they are selected in a manner that the auditor expects to be representative of the population and, thus likely to provide a reasonable basis for conclusions about the population. For example, when the physical representation of the population is a file cabinet drawer of vouchers, a haphazard sample of all vouchers processed for the year 20XX might include any of the vouchers that the auditor pulls from the drawer, regardless of each voucher's size, shape, location, or other physical features.

3.34 As a best practice, the auditor using haphazard selection is usually careful to avoid distorting the sample by selecting, for example, only large, only unusual, only convenient, or only physically small items or by omitting such items as the first or last in the physical representation of the population. The goal is to select a sample without bias. Although haphazard sampling is useful for nonstatistical sampling, it is not appropriate for statistical sampling because it does not allow the auditor to measure the probability of selecting a combination of sampling units.

Block Sampling

3.35 A *block sample* consists of contiguous population items.[4] For example, a block sample from a population of all vouchers processed for the year 20XX might be all vouchers processed on February 3, May 17, and July 19, 20XX. This sample includes only 3 sampling units out of 250 business days because the sampling unit, in this case, is a period of time rather than an individual transaction. A sample with so few blocks is generally not adequate to reach a reasonable audit conclusion. Although a block sample might be designed with enough blocks to minimize this limitation, using such samples might be inefficient. If an auditor decides to use a block sampling technique, he or she

---

[3] When selecting samples on a probability proportional to size basis, such as for monetary unit sampling (MUS), a selection technique known as *cell sampling* reduces or eliminates this problem and can be performed by some computer assisted audit techniques (CAATs). This technique can also be adapted for use in attributes sampling.

[4] A variation of block sampling that can be designed to yield an adequate statistical sampling approach is called *cluster sampling*. The considerations for designing a cluster sample are beyond the scope of this guide. Such guidance can be found in technical references on statistical sampling.

AAG-SAM 3.35

exercises special care to select sufficient blocks to effectively control sampling risk in designing that sample.[5]

3.36 Sometimes auditors will select a number of days from a period and then select a sample of vouchers from those days as a basis for the test. Such a sampling plan actually involves two sampling risks: one related to sampling the days and one related to sampling the items within a day. Sampling expertise may be needed to design a sample that can be expected to be representative to meet the desired overall assurance for the test because these two risks are considered in assessing the sufficiency of the audit evidence.

Determining the Sample Size

3.37 This section discusses the factors that auditors consider when using judgment to determine appropriate sample sizes. Auditors using nonstatistical sampling do not need to quantify these factors; rather, they might consider using estimates in qualitative terms, such as *none*, *few*, or *many*.[6] Appendix A, "Attributes Statistical Sampling Tables," includes additional guidance, along with several tables that can help auditors apply the following discussion to statistical sampling applications.

Considering Sampling Risk in Assessing Controls Effectiveness

3.38 The auditor is concerned with two aspects of sampling risk in performing tests of controls: the risk of erroneously concluding "the controls are more effective than they actually are" (that is, risk of overreliance) and the risk of erroneously concluding "controls are less effective than they actually are" (that is, risk of underreliance).[7] The risk of erroneously concluding controls to be more effective than they actually are is the risk of overreliance on the control caused when the control deviation rate observed in the sample is less than the true deviation rate in the population. Conversely, the risk of erroneously concluding controls to be less effective than they are is the risk of underreliance on the control caused when the control deviation rate in the sample is greater than the true deviation rate in the population.

3.39 The risk of erroneously concluding "controls are less effective than they actually are" (that is, risk of underreliance) relates to the efficiency of the audit. The auditor's assessed level of control risk based on a sample may lead him or her to increase the scope of substantive tests unnecessarily to compensate for the perceived higher level of control risk. Although the audit might be less efficient in this circumstance, it is nevertheless effective. The second aspect of sampling risk in performing tests of controls—the risk of erroneously concluding controls to be more effective than they actually are—relates to the effectiveness of the audit. If the auditor concludes that controls are more effective than they actually are, he or she inappropriately reduces the evidence obtained from substantive procedures. Because the consequences of overreliance are potentially more serious, the following paragraphs relate primarily to that risk.

---

[5] Block samples can be designed as statistical or nonstatistical samples. Sampling specialist assistance may be necessary to design a valid statistical block sample, because there are complexities in computing sampling risk when some blocks are not examined.

[6] Or, depending on the factor, *high*, *moderate*, or *low*.

[7] The term *sampling risk* is defined in paragraph .05 of AU-C section 530, *Audit Sampling* (AICPA, *Professional Standards*).

3.40 Because a test of controls is the primary source of evidence about whether they are operating effectively, the auditor planning to rely on controls generally sets a low risk that the controls will be assessed as more effective than they actually are (that is, a low risk of overreliance).

3.41 There is an inverse relationship between the acceptable risk of overreliance and sample size: the lower the acceptable risk, the larger the sample that is needed. Table 3-1, "Effect on Sample Size of Different Levels of Risk of Overreliance and Tolerable Rate of Deviation," illustrates this relationship. It can be seen that the sample necessary to limit risk to 5 percent is larger than that necessary to limit it to 10 percent. The underlying computations use statistical attributes theory and assume a large population and an expected deviation rate of zero. Instead of quantifying acceptable risk, the auditor may instead characterize it in terms such as *low*, *moderate*, or *high*, but the impact on sample size would be directionally the same.

Table 3-1

Effect on Sample Size of Different Levels of Risk of Overreliance and Tolerable Deviation Rate[1]
(Expected population deviation rate = 0; large population)

Tolerable Rate of Deviation (%)	*Sample Size—10% Risk of Overreliance*	*Sample Size—5% Risk of Overreliance*
10	22	29
5	45	59
1	230	299

[1] Computed using the binomial distribution with sample sizes rounded to the next highest whole number.

3.42 Some auditors find it practical to vary the risk of concluding that controls are more effective than they actually are (that is, risk of overreliance) in response to factors such as the desired level of assurance (or confidence) provided by the test and the availability of other evidence (such as the effective operation of a monitoring complementary or redundant control) to support the test conclusion. An auditor following such a strategy may set a fixed tolerable rate of deviation when designing control test samples, and vary the desired level of assurance or confidence of the test to reflect the other information. For example, absent other information, when the audit strategy calls for (high) reliance on controls, a 90 percent or 95 percent confidence level (for example, 10 percent or 5 percent risk of overreliance) may be used in designing a test. When less assurance is desired, a lower confidence level (for example, 80 percent, 70 percent, 60 percent, or 50 percent) is generally used in designing the test. When additional corroborating evidence of the operation of the control exists, this would also tend to reduce, somewhat, the level of assurance needed from the individual test, depending on the audit evidence available about the factor being considered. For example, a highly effective, documented, and tested management monitoring function may indicate the reasonableness of reducing *high* assurance confidence levels (that is, from 95 percent to 90 percent) on the related controls tests such that a lesser level of assurance is needed from the related test of controls to still achieve a low risk, high assurance result considering the collective testing.

3.43 When planning for tests of controls, some auditors set the tolerable rate of deviation at a fixed rate, and vary the level of assurance or confidence (for instance, the complement of the risk of overreliance) of the test to more easily relate the desired assurance from the test to the audit risk model in paragraphs 4.39–.42 and table 4-2, "Table Relating RMM, Analytical Procedures Risk, and Test of Details (TD) Risk," of this guide, where risk percentages are used to illustrate the risk relationships between the *risks of material misstatement*, including controls and substantive procedures.

3.44 In practice, auditors seeking high controls assurance (that is, low control risk) from a test of a control often set a risk of concluding controls are more effective than they actually are (that is, overreliance) of 10 percent or less. For lesser planned levels of reliance, less assurance is needed. For high risk areas and transactions, such as populations of unusual transactions, nonroutine journal entries, or complex revenue recognition transactions, some auditors increase the desired level of assurance or confidence of controls tests (for example, from 90 percent to 95 percent) in response to these risks.

3.45 Other auditors find it practical to select one level of assurance for all tests of controls (for example, 95 percent) and to assess, for each separate test, a tolerable rate of deviation based on the planned assessed level of control risk. This approach is discussed next. Either approach is acceptable and can lead to adequate sample sizes when properly applied.

Considering Other Evidence in Determining Risk of Concluding Controls are More Effective Than They Actually Are (Overreliance) and Tolerable Rate of Deviation

3.46 In some cases, the auditor may wish to test controls about which evidence from other sources has been obtained. Other sources of evidence include walkthroughs, corroborating inquiries, other evidence about the operation of the control, evidence about the effectiveness of other related controls, competence of personnel, or systems knowledge. In such cases, the auditor may reduce the extent of testing of the control, usually by reducing the level of assurance (increasing the risk of overreliance) or increasing the tolerable rate of deviation used in computing sample size.

Considering the Risk of Concluding Controls are More Effective Than They Actually Are (Overreliance) for Multiple Controls Addressing the Same Control Objective

3.47 The auditor may encounter situations where several redundant or compensating controls address the same control objective or risk. A best practice is for the auditor to first consider the relationship of the controls to the control objective. Depending on that relationship, the auditor may

- test one control at a low level of risk of overreliance, because if that control is operating effectively, the control objective is achieved;
- define the deviation as the failure of both controls to operate on the selected transactions and test at a low level of risk of overreliance;
- test one of the related controls at a low level of risk of overreliance and perform additional testing on other related controls at a higher level of risk of overreliance; or

- test each control at a higher level of risk of overreliance; for example, if each control has a 20 percent risk, the combined risk of the two controls failing is 4 percent if the controls are independent of each other.[8]

Determining the Tolerable Rate of Deviation

3.48 The tolerable rate of deviation for control tests is the maximum rate of deviation from a prescribed control that auditors are willing to accept without altering the planned, assessed level of control risk. Paragraph .05 of AU-C section 530 defines the *tolerable rate of deviation* as

> a rate of deviation set by the auditor in respect of which the auditor seeks to obtain an appropriate level of assurance that the rate of deviation set by the auditor is not exceeded by the actual rate of deviation in the population.

3.49 In determining the tolerable rate of deviation, the auditor usually considers (*a*) the planned assessed level of control risk, and (*b*) the degree of assurance desired by the audit evidence in the sample. Sometimes the auditor specifies a high tolerable rate of deviation because he or she does not require a high level of audit evidence and plans to assess control risk at a higher level. A very high tolerable rate of deviation often implies that the control's operating effectiveness does not significantly reduce the extent of related substantive procedures. In that case, the particular test of controls might be ineffective, and little or no reliance can be placed on the effectiveness of the control.

3.50 In assessing the tolerable rate of deviation, the auditor normally considers that although deviations from pertinent controls increase the *risks of material misstatements* in the accounting records, such deviations do not necessarily always result in misstatements. A recorded disbursement that does not show evidence of an expected approval might, nevertheless, be a transaction that is properly authorized and recorded. Therefore, a tolerable rate of deviation of 5 percent indicates that the test is designed to demonstrate that a control fails no more than 5 percent of the time, and does not necessarily mean that 5 percent of the dollars are misstated. Because not all deviations result in misstatements, auditors usually assess a tolerable rate of deviation for tests of controls that is greater than the comparable tolerable rate of deviation of dollar misstatement.

3.51 When determining a tolerable rate of deviation for a specific control, the auditor normally considers the degree of reliance to be placed on the control and the significance of the control to the audit. The higher the degree of reliance on the control and the greater the significance of the control to the audit, the lower the tolerable rate of deviation.

3.52 There is an inverse relationship between the tolerable rate of deviation and sample size as illustrated in table 3-2, "Effect of Tolerable Rate of Deviation on Sample Size." The table assumes a 10 percent risk of overreliance (90 percent confidence), a large population size, and an expected population deviation rate of zero.

---

[8] The risk of the two independent controls both failing is the combination of the two risks (20 percent multiplied by 20 percent is 4 percent).

Table 3-2
Effect of Tolerable Rate of Deviation on Sample Size[1]
(Assumes a 10 percent risk of overreliance [concluding controls are more effective than they actually are; that is, 90 percent confidence], a large population size, and an expected population deviation rate of 0 percent)

Tolerable Rate of Deviation (%)	Sample Size
3	76
5	45
10	22

[1] Computed using the binomial distribution with sample sizes rounded to the next highest whole number.

3.53 In the usual audit application, when performing tests of controls, the auditor usually is concerned only that the actual rate of deviation in the population does not exceed the tolerable rate of deviation; that is, if, while evaluating the sample results, the auditor finds the sample deviation rate to be less than the tolerable rate of deviation for the population, he or she needs to consider only the risk that such a result might be obtained when the actual deviation rate in the population exceeds the tolerable rate of deviation. The sample-size illustrations in this chapter assume that the sample is designed to measure only the risk that the estimated deviation rate understates the population deviation rate. This is sometimes referred to as an *upper-limit approach*.[9]

3.54 If the auditor finds that the rate of deviation from the prescribed control in the sample plus an allowance for sampling risk (that is, precision) exceeds the tolerable rate of deviation, or that the deviation rate exceeds the expected deviation rate used to design the sample, the auditor typically would conclude that there is an unacceptably high sampling risk. In that case, he or she may increase the assessed level of control risk or consider further whether to rely at all on the control. If statistical sampling has been used, audit software or tables generally are used to calculate the precision of the test (allowance for sampling risk) or the upper limit on the deviation rate.

Considering the Expected Population Deviation Rate

3.55 The auditor estimates the expected population deviation rate by considering such factors as results of the prior year's tests, the design of internal controls, and the control environment. The prior year's results are considered in light of changes in the entity's internal control and changes in personnel.

---

[9] An alternate approach is an interval estimate approach where both an upper and lower limit on the deviation rate is calculated. For a discussion of interval estimates, see Donald Roberts, *Statistical Auditing* (New York: AICPA, 1978): 53.

3.56 There is a direct relationship between the expected population deviation rate and the sample size to be used by the auditor. As the expected population deviation rate approaches the tolerable rate of deviation, the need arises for more precise information from the sample. Therefore, for a given tolerable rate of deviation, the auditor uses a larger sample size as the expected population deviation rate, sometimes referred to as the expected rate of occurrence, increases. Table 3-3, "Relative Effect of the Expected Population Deviation Rate on Sample Size," illustrates the relative effect of the expected population deviation rate on sample size. The table is based on the assumptions of a 5 percent tolerable rate of deviation, a large population size, and a 5 percent risk (95 percent confidence) of overreliance.[10]

Table 3-3

Relative Effect of the Expected Population Deviation Rate on Sample Size[1]
(5 percent tolerable rate of deviation, a large population size, and a 5 percent risk [95 percent confidence] of overreliance)

Expected Population Deviation Rate (%)	Sample Size
0.0*	59
1.0	93
1.5	124
2.0	181
2.5	234

[1] Computed using the binomial distribution with sample sizes rounded to the next highest whole number.

* Some auditors use a sampling approach referred to as *discovery sampling*. Discovery sampling is essentially the same as the approach described in this chapter when the auditor assumes an expected population deviation rate of zero. When used with low risk (high confidence) levels (for example, 1 percent to 2 percent) and low tolerable rates of deviation, discovery sampling has been used in forensic auditing to test for the incidence of rare, unexpected events (such as fraud) in a population.

3.57 The expected population deviation rate would rarely equal or exceed the tolerable rate of deviation. If the auditor believes that the actual deviation rate may be higher than the tolerable rate of deviation, he or she generally omits testing of that control and correspondingly increases the assessed level of control risk.

3.58 The auditor controls the risk of concluding controls are more effective than they actually are by adjusting the sample size for the assessment of the deviation rate he or she expects to find in the population.

---

[10] Large sample sizes, such as 234, are included for illustrative purposes, not to suggest that it would often be efficient to perform tests of controls using such large sample sizes.

Considering the Effect of Population Size

3.59 The size of the population often has little or no effect on the determination of sample size, except in relatively small populations. For example, it is generally appropriate to treat any population of more than 2,000 sampling units as if it were large (for instance, infinite).[11] If the population size is between, for example, 200 and 2,000 sampling units, the population size may have a small effect on the calculation of sample size, depending on the sample parameters. In populations of fewer than 200 items, sample size is reduced by the effect of population size.[12]

3.60 Table 3-4, "Limited Effect of Population Size on Sample Size," illustrates the limited effect of population size on sample size. Computations use statistical theory and assume a 10 percent risk of overreliance (90 percent confidence), a 1 percent expected population deviation rate, and a 10 percent tolerable rate of deviation.

Table 3-4

Limited Effect of Population Size on Sample Size[1]
(Assumes a 10 percent risk of assessing controls as more effective than they actually are—overreliance [90 percent confidence], a 1 percent expected population deviation rate, and a 10 percent tolerable rate of deviation)

Population Size	Sample Size
100	33
200	35
500	37
1,000	37
2,000	38
2,200 or over	38

[1] Computed using the hypergeometric distribution with sample sizes rounded to the next highest whole number.

3.61 Because population size for frequently operating controls has little or no effect on sample size, all other illustrations of sample sizes for tests of controls (except in the next section) assume a large population size.

Small Populations and Infrequently Operating Controls

3.62 Some important controls do not operate frequently, but the auditor may need to test these controls. For example, some controls may be performed only once a year, such as controls over the year-end closing process, and can only be tested once. Other controls are cumulative (for example, a bank reconciliation), so that the auditor may be able to obtain sufficient evidence by

---

[11] Auditors using software that computes sample size and sample results using the hypergeometric distribution will get results that explicitly consider the population size.

[12] Samples not correcting for the smaller population may be inefficient, but still effective.

testing the control at year end (perhaps after doing a walkthrough earlier to understand the control). Still other controls may operate bi-weekly or weekly, such as controls over processing the payroll that may operate 24 or 52 times a year. Such controls may be important, because a significant number of transactions and dollars are controlled by them. The following table provides guidance in the testing of small populations associated with less frequently operating controls.[13] Some auditors applying experience and judgment in the collection of sufficient and appropriate audit evidence have determined that the extent of testing in the following table are reasonable minimums when testing the operating effectiveness of less frequently operating controls. The minimum items to test in this table reflect the assumption that the test may not be a sole source of evidence relating to the control objective in an audit of the financial statements and therefore a higher risk of overreliance is acceptable. In less frequently operating controls, the effect of other sources of evidence is often greater than for more frequently operating controls.[14]

Table 3-5

Testing Operating Effectiveness of Small Populations

Control Frequency and Population Size	Items to Test
Quarterly (4)	2
Monthly (12)	2–4
Semimonthly (24)	3–8
Weekly (52)	5–9

3.63 The number of items to test in table 3-5, "Testing Operating Effectiveness of Small Populations," near the low end of the range may be appropriate for controls reliance in the normal financial statement audit situation. Testing levels in table 3-5 near or even above the upper end of the range presented here may be appropriate in situations when other sources of evidence are less persuasive such as new engagements where there are concerns about the operation of these controls, where controls have changed or where deficiencies had been experienced in the past. When the controls test is the sole source of evidence regarding the effectiveness of these controls, and a specific high level of audit evidence is desired, sampling parameters (for example, risk, tolerable rate of deviation) may be used to determine an appropriate sample size.

Considering a Sequential or a Fixed Sample Size Approach

3.64 Audit samples may be designed using either a fixed sampling plan or a sequential sampling plan. Under a fixed sampling plan, the auditor examines

---

[13] The auditor may need to consider the size of the population by reference to the defined sampling unit. For example, in some cases, the auditor may need to consider the populations from several locations. For example, if there were weekly controls over the occurrence of sales at each of 40 stores, the population of weekly sales test controls would be 2,080 (52 times multiplied by 40), and this would not be a small population.

[14] Some examples of other implicit sources of evidence in an audit of the financial statements include inherent risk assessments, assessments of design and implementation, past experience, walkthroughs, corroborating inquiries, other control testing, knowledge about other balances, competence of personnel, systems knowledge, and so on.

AAG-SAM 3.64

a single sample of a specified size. In *sequential sampling* (sometimes referred to as *stop-or-go sampling*), the sample is taken in several steps, with each step conditional on the results of the previous step. Guidance on sequential sampling plans is included in appendix B, "Sequential Sampling for Tests of Controls," in this guide.

Developing Sample Size Guidelines

3.65 An auditor may establish guidelines for sample sizes for tests of controls based on attributes sampling tables. For example, the sample sizes from the tables in appendix A could form the basis for such guidelines. Some auditors, as a practical and conservative approach, when designing controls tests assume zero deviations initially, and double the sample size if one deviation is found. This approach may not be appropriate for certain controls, such as infrequently occurring controls. Tables and software can be used to more precisely compute sample sizes for specific sampling criteria.

Performing the Sampling Plan

3.66 After the sampling plan has been designed, the auditor selects the sample and examines the selected items to determine whether they contain deviations from the prescribed control.[15] When selecting the sampling units, it is often practical to select several in addition, as extras. If the size of the remaining sample is inadequate for the auditor's objectives, he or she may use the extra sampling units. If the auditor has selected a simple random sample, any additional items used as replacements are generally used in the same order in which the random numbers were generated. The auditor who uses a systematic sampling selection may need to examine all extra selected items.

Voided Documents

3.67 An auditor might select a voided item when selecting a sample. For example, an auditor performing a test of controls related to the entity's vouchers might match random numbers with voucher numbers for the period included in the population; however, a random number might match with a voucher that has been voided. If the auditor obtains evidence that the voucher has been properly voided and does not represent a deviation from the prescribed control, he or she should replace the voided voucher and, if simple random sampling is used, should match a replacement random number with the appropriate voucher.

Unused or Inapplicable Documents

3.68 The auditor's consideration of unused or inapplicable documents is similar to the consideration of voided documents. For example, a sequence of potential voucher numbers might include unused numbers or an intentional omission of certain numbers. If the auditor selects an unused number, he or she would typically obtain evidence that the voucher number actually represents an unused voucher and does not represent a deviation from the control. The auditor then replaces the unused voucher number with an additional voucher number. Sometimes a selected item is inapplicable for a given definition of a

---

[15] Some auditors find it practical to select a single sample for more than one sample objective. This approach is appropriate if the sample size is adequate and selection procedures are appropriate for each of the related sampling objectives.

deviation. For example, a telephone expense selected as part of a sample for which a deviation has been defined as a *transaction not supported by receiving report* may not be expected to be supported by a receiving report. If the auditor has obtained evidence that the transaction is not applicable and does not represent a deviation from the prescribed control, he or she would replace the item with another transaction for testing the control of interest.

Mistakes in Estimating Population Sequences

3.69 If the auditor is using random number sampling to select sampling units, the population size and numbering sequence might be estimated before the transactions have occurred. The most common example of this situation occurs when the auditor has defined the population to include the entire period under audit but plans to perform a portion of the sampling procedure before the end of the period. If the auditor overestimates the population size and numbering sequence, any numbers that are selected as part of the sample and that exceed the actual numbering sequence used are treated as unused documents. Such numbers would be replaced by matching extra random numbers with appropriate documents. If the auditor underestimates the population size and numbering sequence, the auditor will have tested an incomplete physical representation of the population. If this happens, the auditor will generally design additional audit procedures to apply to the items not included in the population.

3.70 In planning and performing an audit sampling procedure, the auditor might encounter the two following special situations.

Stopping the Test Before Completion

3.71 Occasionally the auditor might find a number of deviations in auditing the first part of a sample. As a result, he or she might believe that even if no additional deviations were to be discovered in the remainder of the sample, the results of the sample would not support the planned assessed level of control risk or any reliance on the control being tested. Under these circumstances, the auditor reassesses the level of control risk and considers whether it is appropriate to continue the test.

Inability to Examine Selected Items

3.72 The auditor should perform auditing procedures that are appropriate to achieve the objective of the test of controls on each sampling unit. In some circumstances, performance of the prescribed control being tested is shown only on the selected sample document. If that document cannot be located or if for any other reason the auditor is unable to examine the selected item, he or she considers whether there are alternatives for performing this test on this sample item. In many cases the auditor will probably be unable to use alternative procedures to test whether that control was applied as prescribed. If the auditor is unable to apply the planned audit procedures or appropriate alternative procedures to selected items, he or she should consider selected items to be deviations from the controls for the purpose of evaluating the sample as noted in paragraph .11 of AU-C section 530. In addition, the auditor would typically consider the reasons for this limitation and the effect that such a limitation might have on his or her understanding of internal control and assessment of control risk and audit risk. For example, critical missing documents can be an indicator of fraud, and the auditor may need to consider an appropriate audit response, or, alternatively, whether the missing documentation prevents him or her from concluding on the financial statements as a whole.

Evaluating the Sample Results

3.73 After completing the examination of the sampling units and summarizing the deviations from prescribed controls, the auditor should evaluate the results. Whether the sample is statistical or nonstatistical, the auditor uses judgment in evaluating the results and reaching an overall conclusion.

Calculating the Deviation Rate

3.74 Calculating the deviation rate in the sample involves dividing the number of observed deviations by the sample size. The deviation rate in the sample is the auditor's best estimate[16] of the deviation rate in the population from which it was selected. As a practical matter, deviations may not be present in most samples of controls. Because the purpose of testing is generally to rely on the control, that implies an expectation of effective control operation. Thus, deviations observed in the sample are often important to the auditor's strategy, depending on the deviation rate and reasons for the deviation.

Considering Sampling Risk

3.75 As discussed in chapter 2, sampling risk arises from the possibility that when testing is restricted to a sample, the auditor's conclusions might differ from those he or she would have reached if the test were applied in the same way to all items in the account balance or class of transactions.

3.76 When evaluating a sample for a test of controls, the auditor should evaluate sampling risk. If the estimate of the population deviation rate (the sample deviation rate) is less than the tolerable rate of deviation for the population, the auditor considers the risk that such a result might be obtained even if the true deviation rate for the population exceeds the tolerable rate of deviation for the population. The following is an example of how an auditor might consider sampling risk for tests of controls:

> If the tolerable rate of deviation for a population is 5 percent and no deviations are found in a sample of 60 items, the auditor may conclude that there is an acceptably low risk that the true deviation rate in the population exceeds the tolerable rate of deviation of 5 percent. On the other hand, if the sample includes, for example, two or more deviations (for example, two deficiencies in a sample of 60 items = 3.3 percent), the auditor may conclude that there is an unacceptably higher than planned risk that the rate of deviations in the population may exceed the tolerable rate of deviation of 5 percent.

3.77 If an auditor is performing a statistical sampling application, he or she often uses a table or computer program to assist in measuring the *precision* of the test or the upper limit on control deviations. For example, most computer programs used to evaluate attributes sampling applications calculate an estimate of the upper limit of the possible deviation rate based on the sample size and the sample results at the auditor's specified risk of concluding controls are more effective than they actually are. Table A-3, "Statistical Sampling

---

[16] Also termed the *point estimate* or *direct projection*.

Results Evaluation Table for Tests of Controls—Upper Limits at 5 Percent Risk of Overreliance," and table A-4, "Statistical Sampling Results Evaluation Table for Tests of Controls—Upper Limits at 10 Percent Risk of Overreliance," in appendix A include statistical sampling tables that can help the auditor use professional judgment to evaluate the results of statistical samples for tests of controls at high levels of assurance. The tables may also be useful to auditors using nonstatistical sampling.

3.78 If the auditor is performing a nonstatistical sampling application, sampling risk or precision cannot be measured directly; however, it is generally appropriate for the auditor to conclude that the sample results do not support the planned assessed level of control risk if the rate of deviation identified in the sample exceeds the expected population deviation rate used in designing the sample. When more deviations are encountered than were planned for, the auditor has not met the test objective and there is likely to be an unacceptably high risk that the true deviation rate in the population exceeds the tolerable rate of deviation.[17] In such a circumstance, after considering the reasons for the control deviations and the number of deviations identified, the auditor might conclude it is appropriate to expand the test or perform other tests to include sufficient additional items to reduce the risk to an acceptable level.[18] For example, if a sample of 22 items was sufficient to meet the auditor's objectives, assuming no deviations are expected and one is identified in the sample, to be able to conclude with the same assurance (confidence) as originally planned, the sample needs to be expanded to include many more items. Additional guidance on expanding the sample is provided in this chapter under the heading "Extending the Sample When Control Deviations are Found."

3.79 Rather than testing additional items, however, it is often efficient in a financial statement audit to increase the auditor's assessed level of control risk to the level supported by the results of the original sample and increase the extent of substantive work to reflect the change in the controls assurance. Alternatively, the auditor may decide to place no reliance on the control because the deviation rate found does not support any reliance on the control. For example, if the auditor plans a sample to achieve high assurance expecting one deviation, and two deviations are found in the sample (and no systematic or significant issue is identified when investigating the reason for the deviations), the auditor might be able to conclude with a lower assurance (for example, *moderate* assurance or *limited* assurance) that the control is operating as planned. If a systematic cause is identified, the auditor will typically analyze its effect on controls and potential financial statement misstatement and may conclude that controls reliance at any level is not warranted.

Considering the Qualitative Aspects of the Deviations

3.80 Paragraph .12 of AU-C section 530 states that

---

[17] In accordance with the appendix "Examples of Circumstances That May Be Deficiencies, Significant Deficiencies, or Material Weaknesses" of AU-C section 265, *Communicating Internal Control Related Matters Identified in an Audit* (AICPA, *Professional Standards*), an observed deviation rate that exceeds the number of deviations expected by the auditor in a test of operating effectiveness of a control may indicate a deficiency, significant deficiency, or a material weakness with respect to the operation of the control.

[18] Extending tests introduces additional risks (beyond that measured by the stated risk level) that the auditor might accept a population that should not be accepted.

> the auditor should investigate the nature and cause of any deviations or misstatements identified and evaluate their possible effect on the purpose of the audit procedure and on other areas of the audit

and paragraphs .A22–.A23 of AU-C section 530, respectively, state that

> in analyzing the deviations and misstatements identified, the auditor may observe that many have a common feature (for example, type of transaction, location, product line, or period of time). In such circumstances, the auditor may decide to identify all the items in the population that possess a common feature and extend audit procedures to these items. In addition, such deviations or misstatements may be intentional and may indicate the possibility of fraud.
>
> In addition to the evaluation of the frequency and amounts of monetary misstatements, [paragraph .11 of AU-C] section 450 requires the auditor to consider the qualitative aspects of the misstatements. These include (*a*) the nature and cause of misstatements, such as whether they are differences in principle or application, are errors, or are caused by fraud or are due to misunderstanding of instructions or carelessness, and (*b*) the possible relationship of the misstatements to other phases of the audit. The discovery of fraud requires a broader consideration of possible implications than does the discovery of an error.

3.81 The discovery of fraud will typically elevate the severity of the related control deficiency and the importance of the misstatements to designing other audit procedures.

Extending the Sample When Control Deviations are Found

3.82 The auditor may encounter an unexpected deviation rate in a sample from a population that was expected to be deviation free or to have a low incidence of deviation. Paragraph .12 of AU-C section 530 states that "[t]he auditor should investigate the nature and cause of any deviations or misstatements identified and evaluate their possible effect on the purpose of the audit procedure and on other areas of the audit." In such cases, it is important for the auditor to recognize that the sample is expected to be representative only with respect to the occurrence rate or incidence of deviations, not their nature or cause. An unexpected deviation may be indicative of other deviations in the population. Where the auditor, expecting a negligible or zero deviation rate, selected a small sample, and found a deviation rate slightly higher than expected, then it may be appropriate to extend the sample from that population, but the appropriate extension would not be small. The auditor should first evaluate the nature and cause for the deviation; then, the auditor typically would assess whether, if the sample was extended, the rate of deviations for the combined samples would likely be sufficiently low to support the planned reliance on the control. Extending the sample when the initial sample result was indicative of the true error rate in the population will likely result in further deviations being identified. If there is evidence that the deviation was intentional or could be an indicator of a fraud or there is evidence that conditions could give rise to a systematic or periodic control failure, then extending the test to mitigate the sample findings generally would not be appropriate.

3.83 A properly designed statistical sequential sampling plan (see example in appendix B) or a single stage (fixed) sampling plan designed with an expected deviation rate can be designed in order to draw valid statistical conclusions when deviations are considered to be likely at the outset of the test. Specialist statistical advice may be needed to properly design a custom statistically valid sequential sampling plan.

3.84 When the deviation rate is assessed to be potentially inconclusive or unexpected and extending the test is appropriate, a simple, conservative, rule-of-thumb for expanding single stage samples is to increase the sample size by at least the number of items in the original sample. For example, if a sample of 45 items was sufficient to meet the auditor's control objectives when no deviations were expected, in response to finding one deviation in the first sample of items, the auditor might expand his or her sample by 45 additional items. If no deviations are identified in the additional sample, the combined evidence from the two samples may be sufficient for the auditor to conclude at or near the original level of desired assurance (or risk of concluding controls are more effective than they actually are).[19] Simply adding a few additional items to an initial sample does not have much of an effect on the evaluation of sample results and is generally an inefficient and ineffective procedure. Had the auditor observed two or more deviations when none were expected or planned for, the sample would generally need to be expanded significantly more than the original sample size; often, the auditor will often find it effective and more efficient to not rely on the control than to significantly expand his or her testing of the control. When the auditor uses statistical sampling, a more precise calculation of the needed sample expansion can be made.[20]

Assessing the Potential Magnitude of a Control Deficiency

3.85 If the auditor finds deviations, he or she determines whether they are control deficiencies and, if so, whether those deficiencies are material weaknesses, significant deficiencies, or just deficiencies. One part of this decision is to assess the potential magnitude of each control deficiency.[21] The following discussion focuses on an approach to quantifying the potential magnitude of monetary exposure to misstatement based on control test results. The discussion is limited to the sampling aspects of this approach. AU-C section 265 and AT section 501 include a more robust discussion of quantitative and qualitative factors to consider when assessing the severity of a deficiency in controls.

3.86 Consistent with the guidance in the appendix "Examples of Circumstances That May Be Deficiencies, Significant Deficiencies, or Material Weaknesses" of AU-C section 265, when the auditor identifies control deviations and the deviation rate in the sample exceeds the expected deviation rate used in planning, deficiencies in the design or operating effectiveness of the control are implied. The auditor first understands the nature and cause of the deviations. Then, he or she may apply the following approaches:

- Consider whether other controls, such as redundant or compensating controls, exist that fully or partially mitigate the deficiency

---

[19] This rule of thumb approximates the results of more precise computations that can be made when statistical sampling is applied.

[20] In a statistical calculation, the probability that no deviations will be found in a second sample taken from a population with an unacceptable deviation rate needs to be considered. Such calculations are not considered when using the tables in appendix A, "Attributes Statistical Sampling Tables," which assume a single sample will be drawn and evaluated. The tables in appendix B, "Sequential Sampling for Tests of Controls," are designed for sequential sampling plans.

[21] The issue of assessing likelihood is not fully addressed in this guide.

found in the tested control; if so, understand and test those controls to determine whether the control objective is achieved.
- Assess the likelihood and magnitude of the deficiency, as discussed in the following paragraph.

To apply both approaches at the same time to evaluate a deficiency is usually not appropriate as it would likely understate the severity of the deficiency.[22] However, the auditor could apply the first approach and if not successful in limiting the severity of the deficiency, could apply the upper limit approach (the second approach) as described in the following paragraph.

3.87 When a control does not prevent or detect a misstatement, the auditor would typically conclude, when evaluating the severity of that deficiency, that the related control is likely to fail to prevent or detect misstatements of no less than the magnitude actually observed, and the auditor would then assess the potential magnitude of the control deficiency. The likelihood is generally assessed as high enough to suggest deficiencies when the deviations in the sample exceed the number or proportion of deviations planned for in the sample. If, in a sample of 25 control operations, 1 or more deviations are found, but the sample was expected to have no deviations, then the likelihood criterion is met (assuming the auditor decides not to extend the test). Alternatively, in a sample of 100 control operations where an allowance for 1 deviation was part of the sample design, 1 deviation found in the sample would often indicate that the likelihood criterion has not been met;[23] however, the source and reason for the deviations would be assessed on whether the deviation is a result of any of the factors generally considered to be significant deficiencies or material weaknesses per AU-C section 265; the auditor should consider this when evaluating the severity of the deficiency.

3.88 Control deviations often cannot be equated directly to the potential magnitude of financial misstatement, but in assessing the severity of a deficiency in controls operation, calculating the upper limit on the deviation rate is one way to assist in classifying the deficiency as simply a deficiency, a significant deficiency, or a material weakness. When the auditor is engaged to perform an attestation on the effectiveness of internal controls pursuant to the requirements of AT section 501, such assessments are integral to the purpose of the engagement. For a precise assessment of the dollar impact of control deficiencies, a valid substantive sample would be designed and evaluated. The approach discussed in the following paragraph is a practical adaption to assist auditors in their evaluation of deficiencies.

3.89 A cap on the magnitude of a deficiency may be developed based on an assumption that the upper limit on the deviation rate can be used to roughly estimate the proportion of dollars *exposed* to the control deviation. This estimate, termed *adjusted gross exposure*, may, along with consideration of other quantitative and qualitative factors, assist the auditor in assessing the severity of a deficiency.

3.90 When assessing the significance of a deficiency, qualitative factors are considered in assessing its severity. The qualitative assessment can

---

[22] When the compensating controls are not independent from the control examined, applying both approaches might take "double credit" for mitigating the deficiency, as these approaches are both means to estimate the extent of possible deviation from the observed sample result.

[23] For example, where the sample was designed to allow for one deviation and one deviation was found.

Example

3.91 In a sample of 25 manual control operations from a population of 3,000 control operations, 1 deviation was identified. The sample was designed with an expectation that 0 deviations would be found.

3.92 The sample revealed one deviation (a rate of 4 percent). A statistically based[24] upper limit on the deviation rate can be estimated using software, tables (as illustrated in the following section), or formulas.

Next Steps

3.93 The following illustrates the use of table A-4 in appendix A to this guide:

   a. Locate the sample size (25) along the left column.
   b. Locate the number of deviations (1) along the top row.
   c. Identify the intersection in the body of the table—this is the upper limit (14.7 percent).

Applying the Upper Limit to Measure the Magnitude of Exposure

3.94 The following illustrates how to apply the upper limit to measure the magnitude of exposure:

   a. The sample did not meet its design criteria, so there is probably a higher than desired risk that the control would fail to prevent or detect misstatement. Next, the magnitude of the exposure needs to be assessed.
   b. Gross exposure of the account or process is $5,000,000. This is based on the volume of dollars being processed through the control.
   c. The upper limit on the control deviations, based on the sample result, is 14.7 percent.
   d. The adjusted exposure is $735,000 (14.7 percent × $5,000,000).
   e. The $735,000 adjusted exposure compared to the materiality for the engagement may assist the auditor in evaluating the severity of the control deficiency.

Reaching an Overall Conclusion

3.95 The auditor uses professional judgment to reach an overall conclusion about the effect that the evaluation of the sample results will have on his or her assessed level of control risk, the *risks of material misstatement*, and thus on the nature, timing, and extent of planned substantive procedures. If the sample results, along with other relevant audit evidence, support the planned level of controls reliance, the auditor may not need to modify planned substantive procedures. If the planned assessed level of control reliance is not supported,

---

[24] If the auditor did not select the sample in a random or other statistically valid manner, the result of this evaluation is not *statistical*, but such a computation can still assist auditors in the evaluation of a nonstatistical sample that was expected to be representative of the population.

48 Audit Sampling

the auditor would ordinarily either perform further tests of other controls that could result in supporting the planned level of control reliance or increase the assessed level of control risk and alter the nature, timing, or extent of the planned substantive procedures accordingly.

Documenting the Sampling Procedure

3.96 AU-C section 230, *Audit Documentation* (AICPA, *Professional Standards*), establishes requirements and provides guidance regarding the auditor's responsibility to document audit procedures. Although AU-C section 530 and this guide do not contain a list of specific documentation requirements for audit sampling applications, examples of items that the auditor may document for tests of controls that involve audit sampling include the following:

- A description of the control being tested
- The control objectives related to the sampling application, including the relevant assertions
- The definition of the population and the sampling unit, including how the auditor considered the completeness of the population
- The definition of the deviation condition
- The acceptable risk that controls are more effective than they actually are[25] (or desired confidence or assurance level), the tolerable rate of deviation, and the expected population deviation rate used in the application[26]
- The method of sample size determination
- The method of sample selection
- The selected sample items
- A description of how the sampling procedure was performed
- The evaluation of the sample and the overall conclusion

Paragraph .A14 of AU-C section 230 provides several examples of how an auditor can identify selected sample items in audit documentation.

3.97 The evaluation of the sample and the overall conclusion will generally include the number of deviations found in the sample, the projected deviation rate, an explanation of how the auditor considered sampling risk (for example, the upper limit of the deviation rate for statistical samples), and a determination of whether the sample results support the planned assessed level of control risk. For sequential samples, each step of the sampling plan, including the preliminary evaluation made at the completion of each step, is generally documented. Audit documentation generally will also include the nature of the deviations (if identifiable), the auditor's consideration of the qualitative aspects of the deviations, and the effect of the evaluation on other audit procedures.

3.98 If deficiencies in design or operating effectiveness are found during the tests of controls, the auditor may have reporting responsibilities to management and those charged with governance as noted in AU-C section 265.

---

[25] In other words, the risk of overreliance on controls.

[26] In some instances, sample size inputs such as acceptable risk of overreliance, tolerable rate of deviation, and expected deviation rate are built into firm wide sample size tables. In these instances, reference to firm sample size guidance is sufficient (that is, each team does not need to document inputs that are implicit in the firm's sample size tables).

Chapter 4

Nonstatistical and Statistical Audit Sampling for Substantive Tests of Details

> **⊘ Update 4-1 *Audit*: Clarified Auditing Standards**
>
> The auditing guidance in this guide edition has been conformed to Statement on Auditing Standards (SAS) Nos. 122–125, which were issued in 2011 as part of the Auditing Standards Board's Clarity Project. These clarified SASs are effective for audits of financial statements for periods ending on or after December 15, 2012. Although extensive, the revisions to generally accepted auditing standards resulting from these clarified SASs do not change many of the requirements found in the auditing standards that they supersede.
>
> To assist auditors and financial reporting professionals in making the transition, this guide includes appendix F, "Mapping and Summarization of Changes—Clarified Auditing Standards," which provides a cross reference of the sections in the superseded auditing standards to the applicable sections in the clarified auditing standards and identifies the changes, either substantive or primarily clarifying in nature, that may affect an auditor's practice or methodology relative to the applicable sections of SAS Nos. 122–125. It also summarizes the changes resulting from the requirements of SAS Nos. 122–125.
>
> The preface of this guide and the Financial Reporting Center on www.aicpa.org provide more information on the Clarity Project. Visit www.aicpa.org/sasclarity.

4.01 This chapter introduces the general concepts of audit sampling applicable to both nonstatistical and statistical sampling for substantive tests. Also discussed are guidelines for determining sample size, performing the sampling plan, and evaluating the sample results.

4.02 A purpose of substantive tests of details of transactions and balances is to detect material misstatements in the account balance, transaction class, and disclosure components of the financial statements. An auditor assesses the *risks of material misstatement* and uses a combination of further audit procedures to provide a basis for the opinion about whether the financial statements are materially misstated. When testing the details of an account balance or class of transactions, the auditor might use audit sampling to obtain evidence about the reasonableness of monetary amounts.

4.03 Both statistical and nonstatistical sampling can result in appropriate audit evidence. The auditor may exercise professional judgment in relating the same factors when planning, performing, and evaluating the results of either type of test. Specifically, certain relevant factors (see paragraphs .07–.08, .13, .A13–.A14, and .A27–.A28 of AU-C section 530, *Audit Sampling* [AICPA, *Professional Standards*]) that are equally applicable to both approaches include the following:

- Assessed *risks of material misstatement*
- Characteristics of the population

AAG-SAM 4.03

- Tolerable misstatement
- Expected misstatement
- Audit risk and sampling risk (that is, that actual misstatement exceeds tolerable misstatement)
- Audit evidence obtained from other substantive procedures related to the same assertion
- Selection of a sample that can be expected to be representative
- Projection of the sample results to the population
- Consideration of an allowance for sampling risk (precision)

Determining the Test Objectives

4.04 A sampling plan for substantive tests of details might be designed to (*a*) test the reasonableness of one or more assertions about a financial statement amount (for example, the existence of accounts receivable) or (*b*) make an independent estimate of some amount (for example, the last in, first out [LIFO] index for a LIFO inventory). The first approach, often referred to as *hypothesis testing*, is typically used by an auditor performing a substantive test of details as part of an audit of financial statements. In that case, the auditor accepts an assertion about an amount if it is reasonably correct. The second approach, sometimes referred to as *dollar-value estimation*, is used less frequently by auditors, but might be appropriate when a CPA has been engaged to assist a company in developing independent estimates of quantities or amounts or when the auditor is estimating quantities or amounts as a substantive procedure. For example, a CPA might assist management in estimating the value of LIFO inventory that was previously recorded on a first in, first out basis. Alternatively, a CPA might assist in reconstructing records that were damaged or destroyed. This guide does not provide guidance on the use of sampling if the objective of the application is to develop an independent estimate of quantities or amounts. Furthermore, issues related to independence may be relevant if the auditor develops estimates based on projections from sampling procedures that become the principal basis for the valuation of key accounts in a company's financial statements, and then the auditor opines on the financial statements containing those estimates. Such issues are beyond the scope of this guide.

4.05 It is important that the auditor carefully identifies the characteristic of interest (for example, the misstatement) for the sampling application that is consistent with the audit objective. For example, a characteristic of interest might be defined as differences between the recorded amount and the amount the auditor considers most appropriate, in which case differences related to the characteristic of interest might be called misstatements. Some differences might not involve the characteristic of interest, but may still be important to consider. For example, differences in posting to the correct detail account might not result in misstatement of the aggregate account balance, but may have other audit implications. Also, when the entity has independently identified misstatements and corrected them before the auditor performed procedures on the selected sample items, these items would usually not be considered as misstatements in the sample.[1]

---

[1] However, such information may affect the auditor's assessment of *risk of material misstatement* (RMM) and consequently lead to changes in the nature, timing, or extent of procedures performed.

Defining the Population

4.06 The population consists of the items constituting the account balance or class of transactions of interest subject to audit sampling. It is best practice for the auditor to determine at the beginning of the sampling application that the population from which he or she selects the sample is appropriate for the specific audit objective, because sample results can be projected only to the population from which the sample was selected.[2] For example, an auditor cannot detect understatements of an account that result from omitted items (that is, perform a test of completeness) by sampling only the recorded items. An appropriate plan for detecting such understatements would involve selecting from a source in which the omitted items are included. To illustrate, the auditor might sample (*a*) subsequent cash disbursements for a period of time to test recorded accounts payable for completeness (for instance, understatement) resulting from omitted purchases or (*b*) shipping documents for completeness (for instance, understatement) of sales as evidenced by shipments that were made but not recorded as sales.

4.07 Because the nature of the transactions resulting in debit balances, credit balances, and zero balances typically differ, the audit considerations might also differ because the risks and relevant assertions may differ. Therefore, the auditor usually considers whether the population to be sampled should include all those items together. For example, a retailer's accounts-receivable balance may include both debit and credit balances. The debit balances may result from customer sales on credit, whereas the credit balances might result from advance payments or credit memos and therefore represent liabilities. The audit objectives and assertions for testing those debit and credit balances might be different (for example, the auditor might be more concerned about completeness of credit balances versus existence for the debit balances). If the amount of credit balances is significant, the auditor might find it more effective and efficient to perform separate tests of the debit balances and the credit balances. In that case, the debit and credit balances might be defined as separate populations for the purpose of audit sampling.

Considering the Completeness of the Population

4.08 The auditor typically selects sampling units from a physical representation of the population. If the auditor defines the population as all customer receivable balances as of a specific date, the physical representation might be a trial balance of the customer accounts-receivable subsidiary ledger as of that date.

4.09 The auditor typically considers whether the physical representation includes the entire population. Because the physical representation may be what the auditor actually selects a sample from, any conclusions based on the sample relate only to that physical representation. If the physical representation and the population differ, the auditor might draw erroneous audit conclusions if the auditor projected (extrapolated) the sample results to the entire population.

---

[2] Paragraphs .06 and .13 of AU-C section 530, *Audit Sampling* (AICPA, *Professional Standards*), establish requirements and provide guidance regarding sample design, size, and selection of items for testing and projecting the results of audit sampling, respectively. The definition of *audit sampling* is provided in paragraph .05 of AU-C section 530.

4.10 After footing the physical representation and reconciling it to the population (typically the recorded account balance), the auditor may determine that the physical representation has omitted items in the population that he or she wishes to include in his or her overall evaluation, the auditor would typically select a new physical representation or perform alternative procedures on the items excluded from the physical representation.

Identifying Individually Significant Items

4.11 When planning a sample for a substantive test of details, the auditor uses judgment to determine what items, if any, in an account balance or class of transactions, individually represent relatively high risk of misstatement. These items would often be individually tested and separated from the remainder, which may be sampled. The former category may include items that the auditor judges to be high risk by virtue of size (for example, larger than performance materiality or tolerable misstatement) or risk of misstatement due to error or fraud. In addition, some sampling methods automatically result in items over a certain amount being selected. For example, fixed interval monetary unit sampling results (when material items are not excluded prior to selection) in all items being selected that are greater than or equal to the selection interval.

4.12 Items that the auditor has decided to test 100 percent are not part of the population subject to audit sampling. For example, the auditor might be planning procedures to examine an accounts receivable balance in which 5 large customer balances constitute 75 percent of the account balance. If the auditor examines those balances 100 percent and decides that he or she needs no additional audit evidence for the remaining 25 percent of the account balance because the amounts remaining unexamined are not material and do not represent material risks, or are material and other procedures such as analytical procedures can be effective and will be applied to the amounts, the auditor does not need to use audit sampling, and the examination of that balance would not be covered by AU-C section 530 or this guide; however, if in the auditor's judgment, the remaining items are material in the aggregate and need to be tested using substantive tests of details to fulfill the audit objectives, the auditor might test those remaining items using audit sampling.

Defining the Sampling Unit

4.13 A sampling unit is any of the individual elements that constitute the population. The auditor identifies a sampling unit for a particular audit sampling application. A sampling unit might be a customer account balance, an individual transaction, or an individual entry within a transaction (for example, an individual line item included on a sales invoice).

4.14 The sampling unit depends on the audit objective and the nature of the audit procedures to be applied. For example, if the objective of the sampling application is to test the existence of recorded accounts receivable, the auditor might choose customer balances, customer invoices, or individual items constituting an invoice as the sampling unit. In choosing a sampling unit, the auditor usually considers effectiveness and efficiency in relation to the objective of the test. For example, if the procedure is confirmation of accounts receivable, the auditor may choose a sampling unit that is most likely to elicit a response from the entity's customers. The ease of applying alternative procedures may also be a consideration. For example, if the customer balance is defined as the sampling unit, then the auditor may need to test each individual transaction composing the balance if a customer does not respond.[3] Therefore, it might be more efficient to define the sampling unit as an individual transaction (for example, invoice) composing a customer's accounts-receivable balance.

Choosing an Audit Sampling Technique

4.15 Once the auditor has decided to use audit sampling, either nonstatistical or statistical sampling is appropriate for substantive tests of details. Chapter 2, "The Audit Sampling Process," discusses the general considerations in choosing between a nonstatistical and a statistical sampling approach.

4.16 The most common statistical approaches for substantive testing are classical variables sampling and monetary unit sampling (MUS). Classical variables techniques use normal distribution theory to evaluate the sample results. The MUS approach described in this guide is based on attributes sampling theory.

Selecting the Sample

4.17 In accordance with paragraph .08 of AU-C section 530, the auditor should select the sample in such a way that it can be expected to be representative of the population or the stratum (for instance, without bias) from which it is selected. Auditors using statistical sampling methods follow sample selection approaches appropriate for the statistical technique being used (probability proportional to size [PPS] selection, stratification for classical variables sampling, and so on) that may involve the use of random numbers or the weighting of the probability of an item's selection in proportion to the recorded amount of the item. A nonstatistical sample may be selected using a statistically valid selection technique, or it may be selected using another approach that approximates the selection process for a statistical sample. For example a *haphazard*[4] selection may be designed to approximate a random selection or a PPS selection process. An overview of basic selection methods is presented in chapter 3, "Nonstatistical and Statistical Audit Sampling in Tests of Controls." In addition, PPS selection is discussed in chapter 6, "Monetary Unit Sampling."

4.18 Before selecting the sample, the auditor generally removes high risk items, for instance, those items for which acceptance of some sampling risk is not justified for 100 percent examination. These might include items for which potential misstatements could individually equal or exceed the tolerable misstatement. The auditor may then select the sample directly from the remaining items, use a PPS methodology to select the items, or he or she may

---

[3] Paragraph .12 of AU-C section 505, *External Confirmations* (AICPA, *Professional Standards*), says that "in the case of each nonresponse, the auditor should perform alternative audit procedures to obtain relevant and reliable audit evidence." However, according to paragraph .A26 of AU-C section 505, the omission of alternative procedures may be acceptable (*a*) when the auditor has not identified unusual qualitative factors or systematic characteristics related to the nonresponses, such as that all nonresponses pertain to year-end transactions, and (*b*) when testing for overstatement of amounts, the nonresponses in the aggregate, when projected as 100 percent misstatements to the population and added to the sum of all other unadjusted differences, would not affect the auditor's decision about whether the financial statements are materially misstated.

[4] In this context the term *haphazard* connotes a lack of conscious bias and not carelessness.

stratify the remaining items into groups (strata) and allocate the sample size accordingly.

4.19 As an example of stratification, suppose the accounts-receivable balance includes some large dollar invoices and many small dollar invoices (after excluding the individually significant balances that are examined 100 percent). In that case, the auditor might design the sample to be drawn from two groups: one sample from the group of large dollar invoices and one from the small dollar invoices. table 4-1, "Example of Stratification," shows such groups.

Table 4-1

Example of Stratification

Groups	Items	Recorded Amount
Recorded amount from $100 to $1,000	150	$86,000
Recorded amount up to $100	1,500	$34,000
		$120,000

4.20 The auditor often allocates a portion of the sample to each group. In other words, the sample allocation can more closely approximate a formal stratification plan and be more effective and efficient if the allocation results in a proportionately larger sample size for the large dollar group. For example, after considering the amounts in the population and the risks, the auditor might determine the appropriate sample size to be 60 invoices. If the large dollar group and the small dollar group include recorded amounts of $86,000 and $34,000, respectively, the auditor might select 40 sampling units (in other words, approximately two-thirds, based on a ratio of 86 ÷ 120) from the large dollar group and the remaining 20 sampling units from the small dollar group. The auditor would select the sampling units from each group by any method (for example, haphazard, random, and so on) that can be expected to result in a representative sample of that group.

4.21 Another approach to stratifying the sample or weighting the selection probability proportional to the recorded value of the items[5] is to divide the population into two groups or strata (after excluding those items not subjected to sampling, such as items to be examined 100 percent) with the first group being comprised of items representing approximately half of the sampling population's total monetary value and the second group or stratum representing the other half; then, select half of the sample items from the upper value group and half from the lower value group.

4.22 When the auditor uses stratification approaches such as those just described to select the sample, the sample results are normally separately projected back to each respective stratum and an overall projection is obtained by summing the stratum projections.

---

[5] Sample selection methods that weight the probability of an item's selection to be proportional to its relative size are often appropriate when the primary audit objective is to detect overstatement. See chapter 6, "Monetary Unit Sampling," for additional guidance on when probability proportional to size (PPS) selection may not best meet the auditor's objective.

Determining the Sample Size

4.23 The sample size necessary to provide sufficient audit evidence depends on both the objectives and the efficiency of the sampling methodology. For a given objective, the efficiency of a sample relates to the methodology and its design; one sample is more efficient than another if it can achieve the same objectives with a smaller sample size. In general, careful design can produce more efficient samples.

4.24 If the auditor selects too small a sample, the sample results will not meet the planned objectives. In this case, the auditor ordinarily would perform additional procedures to gather sufficient audit evidence to achieve the planned objectives. If the auditor selects too large a sample, more items than necessary are examined to achieve the planned objectives. In both cases, the audit procedures would often be effective, even though the auditor did not use sampling efficiently. Although audit samples are designed to provide sufficient evidence that an account or population is fairly stated, if misstatements are found, the audit sample may not provide a sufficiently precise estimate for proposing a correcting journal entry (in other words, the uncertainty or precision or statistical bounds around the projected misstatement from the audit sample is too large). Thus, audit samples designed for testing the balance may not be well suited for precise estimation purposes.

4.25 When an audit sample provides evidence that a correcting entry is necessary, the client may decide to perform procedures to determine how much to correct the account, or the client may conduct its own sampling procedures designed to provide a sufficiently precise estimate of the misstatement to support an adjusting journal entry. If the client performs a statistical sample to support an adjusting journal entry, the auditor often performs tests to support the sufficiency and validity of the client's estimation procedure, and may need to obtain the help of a statistical specialist.

4.26 In determining an appropriate sample size for a substantive test of details, the auditor using nonstatistical sampling considers the sampling parameters (for example, risk of misstatement in excess of tolerable misstatement [incorrect acceptance], expected misstatement, and tolerable misstatement) discussed in this chapter, even though he or she might not quantify all of those parameters explicitly. This chapter also includes a table and a risk model that illustrate the relative effects of changes in planning considerations on the determination of sample size.

Considering Variation Within the Population

4.27 Some characteristics, such as the amounts of the individual items in a population, often vary significantly. Accounting populations tend to include a few very large amounts, a number of moderately large amounts, and a large number of small amounts. Auditors frequently consider the variation in a characteristic (for example, recorded amounts, anticipated differences, error distribution within the population, and so on) when they determine an appropriate sample size for a substantive test of details. Auditors often consider the variation of the items' recorded amounts as a means of estimating the variation of the audited amounts of the items in the population.[6] A measure of this

---

[6] Monetary unit sampling (MUS) selection methods do not use this approach (see chapter 6), but the sample is selected with the probability of an item's selection proportional to its size, which some statisticians liken to a form of stratification.

variation, or scatter, is called the *standard deviation*. Auditors using nonstatistical sampling do not need to quantify the expected population standard deviation; rather, they might consider estimating the variation (for example, considering the size of the deviation and its relation to the population) in such qualitative terms as small or large.

4.28 Sample sizes usually decrease as the variation of the sampling characteristic of interest becomes smaller. To reduce the overall variation, a population can be separated, or stratified, into relatively homogeneous groups to reduce the sample size by minimizing the effect of the variation within each group. Sample sizes for unstratified populations with high variation in the sampling characteristic of interest are usually large. To be efficient, stratification is typically based on some characteristic of the items in the population that is expected to reduce variation. When the basis for projecting the sample result is based on misstatements in the sample, the characteristic most relevant to an efficient design of the sample is the variability between the misstatements in the sample, but this statistic is difficult to estimate. Therefore a surrogate, such as recorded amounts, is often used. Other common bases for stratification for substantive procedures include the nature of the controls related to processing the items, or special considerations associated with certain items, such as portions of the population that might be more likely to contain misstatements. Each group into which the population has been subdivided is called a *stratum*. The auditor typically selects separate samples from each stratum and combines the results for all groups in reaching an overall conclusion about the population.[7]

4.29 In addition to affecting sample size, the variation in the population may also affect the approach to selecting the sample by affecting the need for stratification. Auditors using a nonstatistical sampling approach subjectively consider variation within the population. Auditors using a classical variables sampling approach explicitly consider this variability in designing a sampling application. Auditors using MUS do not directly consider this factor because a MUS sample indirectly considers it in the method of sample selection by weighting the probability of an item's selection to be proportional to its size.

4.30 Auditors using a classical variables statistical sampling approach often use a computer program in estimating the variation of a population's audited amounts by measuring the variation of the recorded amounts. Another method of measuring the variation of the items' amounts is to select a pilot sample, which is an initial sample of items in the population. If the auditor is stratifying the population, the pilot sample is usually selected by stratum. The auditor typically performs planned audit procedures on sampling units of the pilot sample and evaluates the pilot sample to gain a better understanding of the variation of both recorded amounts, audited amounts, and misstatements in the population. Although the appropriate size of a pilot sample differs according to the circumstances, it may consist of at least 30–50 sampling units for a large and diverse population.[8] The pilot sample can be designed in a way that allows the auditor to incorporate these items as part of the main sample.

---

[7] Although the projected misstatement results from each stratum are added, the precisions (that is, allowances for sampling risk) related to each stratum are not added, but combined by formula when statistical sampling is used. The formula can be obtained in statistical sampling textbooks. See also Donald Roberts, *Statistical Auditing* (New York: AICPA, 1978): 101. In statistical practice, this approach of separate projection and combination by stratum should be followed.

[8] If the pilot sample is stratified, consideration is also given to selecting a sufficient number of items per stratum.

4.31 Alternatively, the variability of recorded amounts or other applicable characteristics within the population for the prior period may be used to estimate the relevant variability in the current population, provided the underlying processes or expected misstatement conditions have not changed from the prior period. The results of prior years' tests and an adequate understanding of the entity's business and accounting system might provide the auditor with sufficient understanding of the likely variation of amounts in this period without incurring the additional cost of using a pilot sample.

4.32 When adjusting an unstratified variables sample for the lack of stratification, a common range of guidelines call for the sample size to be increased by 10 percent to 50 percent of the computed sample size. In a population with items of about the same amount (after removing items that are insignificant in aggregate and items to be examined 100 percent), such an adjustment may not be necessary. In a population where extreme variability is anticipated in the characteristic of interest (for example, audit differences), the auditor may increase the sample size by 100 percent or more. Typically, stratification of most populations is encouraged to enhance the representativeness of sample selection and the accuracy of the projected sample results. When there is other than a low level of variability in the characteristic of interest (for instance, there are multiple audit differences that vary significantly in size), the auditor may identify this when performing his or her audit procedures if a large sample was taken. In such cases, if the variability used in planning the sample was significantly underestimated, the auditor may need to reconsider the adequacy of the sample to meet the audit objectives.

Determining the Acceptable Level of Risk

4.33 The auditor is concerned with two aspects of sampling risk in performing substantive tests of details: the risk that the sample will lead the auditor to conclude that material misstatement does not exist in the population, when it does (that is, risk of incorrect acceptance) and the risk that the sample will lead the auditor to conclude that material misstatement exists in the population, when it does not (that is, risk of incorrect rejection). The risk of incorrect acceptance and the risk of incorrect rejection are related to the statistical concepts of beta and alpha risk, respectively, as explained in many textbooks on statistical sampling.

The Risk That the Sample Will Lead the Auditor to Conclude That Material Misstatement Does Not Exist in the Population, When it Does (Incorrect Acceptance)

4.34 The risk of incorrect acceptance is the risk that the sample supports the conclusion that the recorded account balance is not materially misstated when it is materially misstated. In assessing an acceptable level of the risk of incorrect acceptance, the auditor typically considers (*a*) the level of audit risk that he or she is willing to accept, (*b*) the assessed *risks of material misstatement* (considering both inherent and control risks), and (*c*) the detection risk for further audit procedures directed toward the same specific audit objectives or financial statement assertions, including further tests of controls, analytical procedures, and substantive tests of details not involving audit sampling.

4.35 For a particular population, audit risk is the risk that there is monetary misstatement greater than tolerable misstatement and that the auditor fails to detect it. Auditors use professional judgment in determining the acceptable audit risk for a particular account balance or class of transactions and related assertions, after considering such factors as the *risks of material misstatement* in the financial statements, the cost to reduce the risk, and the effect of the potential misstatement on the use and understanding of the financial statements.

4.36 The extent of substantive procedures to obtain sufficient audit evidence varies with the auditor's assessed *risks of material misstatement*. Also, the extent of the audit evidence required from a particular substantive procedure varies with the risk that other substantive procedures will fail to detect a material misstatement of the assertion being audited. Paragraphs .A46–.A48 of AU-C section 200, *Overall Objectives of the Independent Auditor and the Conduct of an Audit in Accordance With Generally Accepted Auditing Standards* (AICPA, *Professional Standards*), provides guidance on detection risk and its effect on the auditor's substantive procedures.

4.37 The combination of the auditor's *risk of material misstatement* and consideration of the results of further audit procedures provide the basis for the auditor's opinion. The lower the *risk of material misstatement* or the greater the reliance on other tests directed toward the same specific audit objective (or assertion), the greater the allowable risk of incorrect acceptance (and the lower the desired level of confidence) for the substantive test of details being planned, and, thus, the smaller the required sample size for the substantive test of details. For example, if the auditor assesses the *risk of material misstatement* to be high and performs no other substantive test of details to achieve the same objectives, he or she should plan to achieve a low risk of incorrect acceptance (a high level of desired confidence) for the substantive test of details. Thus, the auditor would select a larger sample for the test of details when the *risk of material misstatement* is high than when the *risk of material misstatement* was low.

4.38 A planning model expressing the general relationship of audit risk to the assessed *risks of material misstatement* and detection risk is described in paragraph .A46 of AU-C section 200.

The Audit Risk Model

4.39 The following risk model[9] illustrates a quantitative method of enhancing the auditor's understanding of the relative effect of the *risks of material misstatement* (RMM) and analytical procedures risk on the size of samples for substantive tests of details.[10] Further discussion of the risk model elements (for example, inherent risk, control risk, RMM, and detection risk) is found in AU-C section 200.

4.40 There is no requirement that the auditor express audit judgments in terms of risk percentages or make computations of audit risk. The model is provided to illustrate the relative effect of different planning considerations on sample size; it is intended as an aid and not a substitute for professional judgment. When using this model, the auditor still applies professional judgment

---

[9] This risk model was previously published as the appendix "Relating the Risk of Incorrect Acceptance for a Substantive Test of Details to Other Sources of Audit Assurance" in AU section 350, *Audit Sampling* (AICPA, *Professional Standards*). A form of this risk model was originally introduced in 1981 in table 2 of Statement on Auditing Standards (SAS) No. 39, *Audit Sampling*, and revised in 2005 by SAS No. 111, *Audit Sampling* (AICPA, *Professional Standards*, AU sec. 350).

[10] The table assumes the items to be examined 100 percent have already been removed from the population.

Nonstatistical and Statistical Audit Sampling for Substantive Tests of Details

in assessing all the factors to be used in designing the test of details, and, in addition, assesses

- the *risks of material misstatement* (or inherent and control risk or RMM); and
- the risk that other substantive procedures (for example, analytical procedures [AP]) will fail to detect a material misstatement.

Table 4-2

Table Relating RMM, Analytical Procedures Risk, and Test of Details (TD) Risk
Allowable Risk of Incorrect Acceptance (TD) for Various Assessments of RMM and AP (for Audit Risk [AR] = .05)

Auditor's subjective assessment of *risk of material misstatement.*	Auditor's subjective assessment of risk that substantive analytical procedures and other relevant substantive procedures might fail to detect aggregate misstatements equal to tolerable misstatement.			
RMM	AP			
	10%	30%	50%	100%
	TD			
10%	*	*	*	50%
30%	*	55%	33%	16%
50%	*	33%	20%	10%
100%	50%	16%	10%	5%

* The allowable level of AR of 5 percent exceeds the product of RMM and AP, and, thus, the planned test of details may not be necessary unless specified by regulation or other standards (for example, confirmation or inventory observation procedures).

Note: The table entries for TD are computed from the illustrated model: TD equals AR ÷ (RMM × AP). For example, for RMM = .50, AP = .30, TD = .05 ÷ (.50 × .30) or .33 (equals 33%).

4.41 For example, suppose the auditor using the table 4-2, "Table Relating RMM, Analytical Procedures Risk, and Test of Details (TD) Risk," relationships assesses the *risks of material misstatement* (for example, 50 percent) and the risk that analytical procedures might not detect material misstatement (for example, 50 percent). Table 4-2 indicates that a 20 percent risk (in other words, 80 percent confidence level) for a related test of details is appropriate.[11] Some auditors express these risks using terms like *high*, *moderate*, and *low* rather than using estimates of risk percentages.

---

[11] The auditor can calculate the acceptable test of details risk for any combination of risks by using the formula: Audit Risk (AR) = RMM × Analytical Procedures (AP) Risk × Test of Details Risk (TD) and solving for the test of details risk. Audit risk is illustrated as being set at 5 percent.

AAG-SAM 4.41

4.42 When the auditor has performed only an assessment of design and implementation of internal controls and assessed the design as effective and has obtained evidence that the controls have been implemented, the auditor might accept a slightly higher risk of incorrect acceptance (lower confidence level) for substantive tests of details than had the design or implementation of controls been assessed as ineffective.[12]

The Risk That the Sample Will Lead the Auditor to Conclude that Material Misstatement Exists in the Population, When it Does Not (Incorrect Rejection)

4.43 The risk of incorrect rejection is related to the efficiency of the audit. For example, if the auditor's evaluation of a sample leads him or her to an initially erroneous conclusion that a balance is materially misstated when it is not, the consideration of other audit evidence and performance of additional audit procedures would ordinarily lead the auditor to the correct conclusion. When auditors decide to limit the risk of incorrect rejection, they typically increase the sample size for the (substantive) test of details; they also decrease the risk that they might incur costs for performing additional procedures to resolve differences between a correct recorded amount and an erroneous estimate resulting from an inadequately controlled risk of incorrect rejection. Although the audit might be less efficient in this circumstance, it is effective. Some auditors have determined that the larger sample sizes required to limit the risk of incorrect rejection across all sampling applications is too costly, so these auditors do not usually design samples to limit the risk of incorrect rejection. Rather, these auditors have decided it is better to incur the costs of performing additional procedures in those situations when they find a higher amount of misstatement than expected. In other cases, the auditor decides whether and how to address the risk of incorrect rejection on a sample by sample basis.

4.44 Some auditors provide some protection against the risk of incorrect rejection by conservatively estimating the amount of expected misstatement when planning the sample, thereby increasing the sample size. Other auditors may add an additional percentage of items (for example, 10 percent) to the computed sample size; however, these methods do not specifically control how much protection is obtained.

4.45 Other auditors decide whether and how to address the risk of incorrect rejection on a sample by sample basis. These auditors may limit the risk of incorrect rejection when the extension of the original sample, after sample evaluation, will be extremely costly in terms of additional sampling cost or the timing of the findings (for example, it is not physically practical to revisit a site to extend the work [such as when visiting remote locations], or the time required to perform additional tests may significantly delay financial reporting).

4.46 In very low expected misstatement populations, when the assurance desired from the sample is low, and when the client will adjust for some projected, as well as factual, misstatement, the risk of incorrect rejection is less important when planning the sample because the inefficiencies of this risk are less in such situations.

---

[12] To place significant reliance on controls, paragraph .08 of AU-C section 330, *Performing Audit Procedures in Response to Assessed Risks and Evaluating the Audit Evidence Obtained* (AICPA, *Professional Standards*), indicates that the auditor should assess design and implementation *and* test the operating effectiveness of the control.

4.47 The auditor is usually more concerned with the risk of incorrect rejection when planning a sampling application for substantive testing than with the risk of underreliance when planning a sampling application for a test of controls, although both risks have efficiency considerations. If the sample results for a test of controls do not support the auditor's planned assessed level of control risk, the auditor generally performs additional tests of controls to support the planned assessed level of control risk, or increases the planned assessed level of substantive testing in response to the test results. Because an alternative audit approach is readily available, the inconvenience to the auditor and the entity resulting from the risk of underreliance is usually relatively small; however, if the sample results for a (substantive) test of details support the conclusion that the recorded account balance or class of transactions is materially misstated when it might not be, the alternative approaches available to the auditor might be more costly, and become known only at a critical point in the summarization of the audit findings. In most cases, the auditor would have further discussions with the entity's personnel and may perform additional audit procedures. The cost of this additional work might be substantial and the timing may also be very impractical. Further consideration of the risk of incorrect rejection is discussed in chapters 6–7.

Considering Tolerable Misstatement

4.48 Tolerable misstatement, as defined in paragraph .05 of AU-C section 530, is

> a monetary amount set by the auditor in respect of which the auditor seeks to obtain an appropriate level of assurance that the monetary amount set by the auditor is not exceeded by the actual misstatement in the population.

4.49 When planning a sample for a substantive test of details, the auditor typically considers how much monetary misstatement in the tested assertion may exist, when combined with misstatements that may be found in other tests in this and other accounts without causing the financial statements to be materially misstated. The auditor usually then designs the test to provide sufficient assurance that the population does not contain misstatements greater than this amount.

Performance Materiality and Tolerable Misstatement

4.50 Tolerable misstatement is related to the auditor's required assessment of performance materiality, which is used for purposes of assessing the risks of material misstatement and determining the nature timing and extent of further audit procedures. Tolerable misstatement for an account, balance, or class of transactions is normally set at, or less than performance materiality. Tolerable misstatement allows for a meaningful comparison of the results of procedures in an account, balance, or class of transactions with the related performance materiality, and for the aggregation of the results of tests of the accounts to compare to materiality.[13] Both performance materiality and tolerable misstatement can be used to make a provision for possible misstatements that might exist in the financial statements, but were not detected by the

---

[13] Paragraph .11 of AU-C section 320, *Materiality in Planning and Performing an Audit* (AICPA, *Professional Standards*), states that the auditor should determine performance materiality for purposes of assessing the risks of material misstatement and determining the nature, timing, and extent of further audit procedures.

audit procedures. For a given risk of incorrect acceptance, sample sizes tend to increase directly as performance materiality, tolerable misstatement, or both, decrease. The concept of performance materiality is explained in paragraph .A14 of AU-C section 320, *Materiality in Planning and Performing an Audit* (AICPA, *Professional Standards*), which states that

> planning the audit solely to detect individual material misstatements overlooks the fact that the aggregate of individually immaterial misstatements may cause the financial statements to be materially misstated and leaves no margin for possible undetected misstatements. Performance materiality (which, as defined, is one or more amounts) is set to reduce to an appropriately low level the probability that the aggregate of uncorrected and undetected misstatements in the financial statements exceeds materiality for the financial statements as a whole. Similarly, performance materiality relating to a materiality level determined for a particular class of transactions, account balance, or disclosure is set to reduce to an appropriately low level the probability that the aggregate of uncorrected and undetected misstatements in that particular class of transactions, account balance, or disclosure exceeds the materiality level for that particular class of transactions, account balance, or disclosure. The determination of performance materiality is not a simple mechanical calculation and involves the exercise of professional judgment. It is affected by the auditor's understanding of the entity, updated during the performance of the risk assessment procedures, and the nature and extent of misstatements identified in previous audits and, thereby, the auditor's expectations regarding misstatements in the current period.

4.51 Paragraph .A6 of AU-C section 530 states that "*Tolerable misstatement* is the application of performance materiality to a particular sampling procedure. Tolerable misstatement may be the same amount or an amount smaller than performance materiality (for example, when the population from which the sample is selected is smaller than the account balance)."

4.52 Tolerable misstatement may be thought of as an extension of the concept of performance materiality applied at the test level to accounts, balances, or classes of transactions. As performance materiality is less than materiality to provide for the combination of various audit areas, similarly tolerable misstatement may be less than performance materiality when there is a need to incorporate some additional factors that may not have been fully considered when setting performance materiality such as differing population characteristics (for example, expected misstatement) in certain sample tests or the combination of test results from various samples or estimating procedures in an audit area before comparison to performance materiality. How tolerable misstatement is set relative to performance materiality depends on what specific factors were considered when determining performance materiality and what additional factors, if any, need to be considered when determining tolerable misstatement. For example, if several samples and estimation procedures are being used in auditing the inventory account, and the expectation of misstatement in these tests was different from sample to sample and from the expectations of misstatement in tests from other audit accounts or areas, and these considerations were not factored into setting the performance materiality for the various accounts (see table 4-3, "Factors to Consider in Setting Performance Materiality (PM) at the Engagement Level and Tolerable Misstatement (TM) at the Test Level"), then tolerable misstatement might be set below performance

Nonstatistical and Statistical Audit Sampling for Substantive Tests of Details **63**

materiality for some tests to reflect these differing characteristics. Had all the factors that influence performance materiality and tolerable misstatement been considered in determining the performance materiality amount, then tolerable misstatement might be set at performance materiality. Also, as stated in AU-C section 530, tolerable misstatement may be set lower than performance materiality when the sample is performed on only a portion of the account.

4.53 Auditors may consider specific factors when determining performance materiality or whether to set tolerable misstatement at the same amount or less than performance materiality. These factors can apply to setting either or both performance materiality and tolerable misstatement. Application of these factors can result in a wide range of possible relationships between performance materiality, tolerable misstatement and materiality, but the determination of the relationship is a judgment based on the circumstances of the application. Some auditors may consider some or all of these factors when setting performance materiality. For efficiency reasons, others may consider some of the factors when setting performance materiality and consider other factors when determining tolerable misstatement for specific accounts, balances, or transactions. This latter approach adjusts the extent of sampling to respond to the differences in certain factors between tests and accounts (for example, expected misstatement, resistance to correction, and the aggregation of various tests before comparison to performance materiality).[14] Appendix L, "Matters to Consider in Determining Performance Materiality," of the AICPA Audit Guide *Assessing and Responding to Audit Risk in a Financial Statement Audit* enumerates the factors normally considered in setting performance materiality. The following table 4-3 expands that appendix to include factors that may additionally be considered at the tolerable misstatement (test) level.

Table 4-3
Factors to Consider in Setting Performance Materiality (PM) at the Engagement Level and Tolerable Misstatement (TM) at the Test Level

Factors to Consider in Setting PM and TM Misstatement	Conditions Leading to a PM and TM Much Lower Than Materiality	Conditions Leading to a PM and TM Closer to Materiality	Comments
Expected total amount of factual and projected misstatements (based on past significant misstatements and other factors)	A greater number of misstatements	A lesser number of misstatements	The allowance for undetected misstatements is typically greater when more misstatements are expected.

(continued)

---

[14] When all the factors are considered in the determination of performance materiality, then any differing characteristics between the populations tested in different samples may be "averaged" and the individual tests may not be responsive to any differences in population characteristics, possibly affecting sampling efficiency.

AAG-SAM 4.53

Factors to Consider in Setting PM and TM Misstatement	Conditions Leading to a PM and TM Much Lower Than Materiality	Conditions Leading to a PM and TM Closer to Materiality	Comments
Management's attitude toward proposed adjustments	Management is generally resistant to adjustments	Management is open to considering adjustments and usually corrects all factual misstatements and many projected misstatements	More adjustments of factual and projected misstatements will lessen the amount needed to allow for undetected misstatements.
Number of accounts or tests in an audit area where amounts will be subject to estimation and will not be able to be determined with precision	A significant number of accounts (PM) or tests or estimates in an account or area (TM)	One or a few accounts (PM) or one or a few tests or estimates in an audit area (TM)	A greater allowance for undetected misstatements is needed when there are more accounts (or tests) that are subject to estimation procedures.
Locations, subsidiaries, or samples *within an account* where separate procedures are applied for each location but that will be aggregated in reaching audit conclusions	A significant number of locations, subsidiaries, or samples within an account (TM, PM, or both)	One or a few locations, subsidiaries, or samples within an account (TM, PM, or both)	A greater allowance for undetected misstatements is needed due to the imprecision of many samples (TM).
Portion of an account or area included in a test	A smaller portion of the account is being tested (TM)	A large portion or the whole account is being tested (TM)	At the individual test level, the tolerable misstatement is generally reduced when only a portion of the account is tested.

4.54 As practical guidance, across many engagements, setting the relationship of tolerable misstatement ranges of 50 percent to 75 percent of materiality has often been shown to be adequate to allow for the factors influencing the performance materiality and tolerable misstatement amount (for example, the factors noted in table 4-3). The guideline in paragraph 2.26 of the AICPA Audit Guide *Assessing and Responding to Audit Risk in a Financial Statement Audit* relates the factors in appendix L of that same guide to setting performance materiality. When some factors are considered in setting performance materiality and additional factors are considered when setting the tolerable misstatement level, then these same guidelines (for example, 50 percent to 75 percent) can be applied when relating tolerable misstatement to materiality.

4.55 According to paragraph .09 of AU-C section 320, *performance materiality* is the amount or amounts set by the auditor at less than materiality for the financial statements as a whole to reduce to an appropriately low level the probability that the aggregate of uncorrected misstatements exceeds materiality for the financial statements as a whole. Accordingly, setting performance materiality involves the use of professional judgment. Performance materiality can be the same or different for different accounts, and some of the factors in table 4-3 may have already been considered in making this judgment (for example, number of locations, number of accounts). Other factors, such as the number of sampling and estimating procedures performed in an account or the portion of an account included in the test, may be considered when determining the tolerable misstatement amount. Additionally, the reluctance of management to make adjustments and the amount of misstatement expected to be found in a specific audit sample may not have been considered in determining performance materiality because these factors might differ greatly from account to account and sample to sample. To vary the level of testing in accounts, to be responsive to some of these population differences, these factors might be considered when setting tolerable misstatement. Consequently, the judgment regarding the level of tolerable misstatement for samples within a balance or account may depend on the approach to setting performance materiality and the factors that were incorporated into that judgment. As a result, it is impractical to set general[15] guidelines for how to set these parameters separately in relation to each other. In general, the relationship of tolerable misstatement to materiality is the product of the two assessments.[16] Although some auditors set a single overall performance materiality (or tolerable misstatement) relationship for all accounts (or tests), others may vary the relationship somewhat to reflect risk and efficiency characteristics. Whether the relationship between performance materiality, tolerable misstatement and materiality is or is not varied between accounts, the overall audit risk and adequacy of the allowance for sampling risk is still to be determined for the aggregate of samples.[17]

4.56 Note, however, that such planning calculations can imply a degree of testing precision that is not actually attainable in the audit, because many

---

[15] It may be possible in a completely statistical sampling setting to determine sample sizes and the required tolerable misstatement and performance materiality levels to meet specific confidence and precision parameters.

[16] When both performance materiality for an audit area is set lower than materiality and tolerable misstatement for tests in that area is set lower than performance materiality, the combined impact on the relationship between tolerable misstatement and materiality is the product of the two assessments (for example, 80 percent PM × 90 percent TM = TM is 72 percent of materiality).

[17] For a theoretical development of this concept see Saurav Dutta and Lynford Graham, "Considering Multiple Materialities for Account Combinations in Audit Planning and Evaluation," *Journal of Accounting, Auditing and Finance* (Spring 1998): 151–171. Theoretically, the most efficient strategy for setting and balancing the tolerable misstatement for individual accounts considers both the risks and costs of performing procedures in the accounts.

AAG-SAM 4.56

of the parameters of the population (for example, standard deviations and expected misstatements) often are estimated and are not known with certainty. Additionally, audit samples are typically not the sole source of substantive evidence regarding assertions, accounts, and balances. Ordinarily, substantive analytical procedures such as using expectations based on turnover, ratio, trend analyses, and agings, or other audit tests, will also provide evidence regarding the reported balances. When other substantive and control procedures are applied, they too contribute to reducing the risk in the various accounts, but direct measurements of these contributions are difficult, as statistical measures of their risk and precision characteristics may not be determinable; however, when the contributions of these other procedures can be measured, it would tend to decrease the need to reduce the performance materiality and tolerable misstatement measures relative to materiality.[18]

Observations and Suggestions

The benchmark of relating tolerable misstatement to materiality is based on prior sampling guidance and experience. Because of the different ways auditors might plan for performance materiality and the many similar factors influencing performance materiality and tolerable misstatement, the rule of thumb relating tolerable misstatement to materiality is suggested when all of the factors are not fully incorporated into determining performance materiality. Performance materiality is a new concept, and is positioned between materiality at the engagement level and tolerable misstatement at the sample level. Different approaches to setting performance materiality will have an effect on the required tolerable misstatement to enable the auditor to aggregate various procedures at the end of the audit and opine based on obtaining a level of audit assurance that the financial statements are free of material misstatement.

Although a rule of thumb can be helpful in relating tolerable misstatement to performance materiality and then materiality, the setting of tolerable misstatement relative to performance materiality considers those factors in table 4-3 that were not incorporated into setting the performance materiality. As an example, suppose two accounts have the same recorded amount and the same quantitative relationship of the recorded balance to performance materiality (for example, recorded amount/performance materiality is 70 percent). In one account, a number of samples and estimation procedures are planned (for example, a sample for existence, a sample for pricing, an obsolescence estimate) and some misstatement that may not be corrected is expected. In the second account, the auditor plans to reperform the depreciation calculations for scheduled assets and may only need to perform one sample to test the calculations for new additions to the account, and no misstatement is expected. Had all the relevant factors been incorporated into the determination of performance materiality, then tolerable misstatement may be set at or near performance materiality. However, if these differing account-level characteristics were not factored into determining the performance materiality assessments for each account at the outset, then the first account would usually set tolerable misstatement at less than performance materiality, to allow for the expected

---

[18] This paragraph relates to the third factor in table 4-3, "Factors to Consider in Setting Performance Materiality (PM) at the Engagement Level and Tolerable Misstatement (TM) at the Test Level."

misstatement and provide for an allowance for sampling risk for the combination of various sampling and estimation procedures within the account or audit area. In the second account, tolerable misstatement for the planned procedure (a single sample) might be set at the performance materiality amount.

Special Topics Related to Determining Populations and Tolerable Misstatement

4.57 *Tolerable misstatement for reclassifications.* Most audit samples are designed to simultaneously gather evidence about assertions in both the income statement and the balance sheet. In most cases, it is not appropriate to set tolerable misstatement greater than performance materiality or materiality.[19] However, in limited situations, (*a*) the audit evidence obtained from other audit procedures may be sufficient to conclude that a potential misstatement of an income statement or balance sheet account could result only in a reclassification that would not affect net income and its classifications or significant balance sheet classifications; and (*b*) any potential misstatements identified by a planned procedure would not affect other significant measures of financial performance (for example, current ratio; gross margin; operating income; earnings before interest, taxes, depreciation and amortization; covenant thresholds; and so on). When these conditions are present, then the auditor may use a tolerable misstatement based on a larger threshold for reclassifications.

4.58 *The use of gross margin in sample planning.* Auditors typically define a population as the recorded amount of all items composing the account balance or class of transactions being tested. Revenue and cost of sales are most often regarded as two separate classes of transactions and therefore two populations for sampling purposes. Accordingly, it is usually inappropriate to seek reduced sample sizes by planning an audit sample to test revenue or cost of sales using a single net population defined as the *gross margin*. This approach may incorrectly assume that misstatements of revenue are always offset by misstatements in cost of sales or vice versa. For example, as a result of fraud, fictitious revenues may be recorded without any matching cost. As a further example, cut-off errors might represent misstatements of either revenue or cost of sales but not necessarily both. Samples designed assuming only a gross margin population is *at risk* may be too small to provide the desired level of assurance that these and similar sources of misstatement would be detected.

4.59 *Designing samples to address assertions.* Normally, the amount *at risk* in a population is the amount that is exposed to misstatement relative to the assertion of interest. In relation to the existence assertion, this is usually the total amount of the sample item's value. Although it is usually inappropriate to regard anything less than the gross amount to be at risk across all assertions, in some limited and unusual situations, the auditor may have obtained sufficient appropriate audit evidence regarding fraud risk, existence and occurrence, and completeness of the recorded balance, but not have sufficient appropriate evidence regarding another assertion (for example, accuracy). In such circumstances, the auditor may devise appropriate procedures to obtain the desired level of assurance on a specific assertion such as accuracy.

---

[19] Normally, the auditor would relate materiality, performance materiality, and tolerable misstatement to the items affecting net income when these measures are relevant, because income measures of materiality are often smaller than measures based on balance sheet amounts such as for reclassifications.

For example, if the auditor of a financial institution has obtained substantial assurance regarding existence of loans through regulatory or internal audit confirmation procedures and is designing the sample to address primarily the interest impact of the terms of the loan that were not confirmed in the prior tests, then such an approach may be appropriate.

Considering the Expected Amount of Misstatement

4.60 In determining the sample size, the auditor typically considers the total amount of misstatement he or she expects to find in the population. In general, as the expected amount of misstatement approaches the tolerable misstatement, there is a need for more precise information from the sample. Therefore, the auditor would usually expect this to result in a larger sample size as the expected amount of misstatement increases.

4.61 The auditor may assess the expected amount of misstatement on the basis of his or her professional judgment after considering such factors as the entity's business and risks, the results of prior years' tests of the account balance or class of transactions, the results of any pilot sample, the results of any related substantive procedures, and the results of any tests of the related controls or changes to the controls during the year.

Considering the Effect of Population Size

4.62 The number of items (for example, invoices) in a large population often has little effect on the determination of an appropriate sample size for substantive tests of details; however, when the population consists of a small number of very significant, but not individually material items, the concepts of audit sampling can be difficult to apply, and the auditor may need to consult with a sampling specialist when designing procedures in such circumstances. If an auditor wants to apply audit sampling to a small population, the sample sizes produced by some sampling methods that do not consider population size may be too large for the purpose, although still effective. Some auditors have applied statistical factors or formulas to resize such samples. When applying classical variables sampling using either mean per unit or difference estimation, the auditor needs an estimate of population size to accurately estimate projected misstatement and the allowance for sampling risk in dollars. When using some methods of MUS, the auditor needs to know the total recorded dollar amount of the population, for example, to select the sample and project the sample result.

Relating the Factors to Determine the Sample Size

4.63 An understanding of the relative effects of various planning considerations on sample size is useful in designing an efficient sampling application. The auditor uses professional judgment and experience in considering those factors to determine a sample size. Table 4-4, "Factors Influencing Sample Sizes for a Substantive Test of Details in Sample Planning," summarizes the effects of various factors on sample sizes for substantive tests of details. The table is provided only to illustrate the relative effect of different planning considerations on sample size; it is not intended as a substitute for professional judgment.

Table 4-4
Factors Influencing Sample Sizes for a Substantive Test of Details in Sample Planning

Factor	Conditions Leading to:		Related Factor for Substantive Sample Planning
	Smaller Sample Size	Larger Sample Size	
a. Assessment of inherent risk	Low assessed level of inherent risk	High assessed level of inherent risk	Allowable risk of incorrect acceptance
b. Assessment of control risk	Low assessed level of control risk	High assessed level of control risk	Allowable risk of incorrect acceptance
c. Assessment of risk related to other substantive procedures directed at the same assertion (including substantive analytical procedures and other relevant substantive procedures)	Low assessment of risk associated with other relevant substantive procedures	High assessment of risk associated with other relevant substantive procedures	Allowable risk of incorrect acceptance
d. Measure of tolerable misstatement for a specific account	Larger measure of tolerable misstatement	Smaller measure of tolerable misstatement	Tolerable misstatement
e. Expected size and frequency of misstatements, or the estimated variance of the population	Smaller misstatements or lower frequency, or smaller population variance	Larger misstatements, higher frequency, or larger population variance	Assessment of population characteristics
f. Number of items in the population	Virtually no effect on sample size unless population is very small[20]		

4.64 Paragraph .A14 of AU-C section 530 clarifies that sample sizes of statistical and nonstatistical samples ordinarily would be comparable in similar situations, by stating that

---

[20] Some statistical substantive sampling techniques and formulas do consider population size in the determination of sample size, but in most cases the number of logical units in the population will not affect the resulting sample size much, unless the population is very small.

an auditor who applies nonstatistical sampling exercises professional judgment to relate the same factors used in statistical sampling in determining the appropriate sample size. Ordinarily, this would result in a sample size comparable with the sample size resulting from an efficient and effectively designed statistical sample, considering the same sampling parameters. This guidance does not suggest that the auditor using nonstatistical sampling also compute a corresponding sample size using an appropriate statistical technique.

4.65 In the preceding, the parameters or factors may include RMM, sampling risk, performance materiality and tolerable misstatement, and expected misstatement.

4.66 Even though sample sizes between statistical and nonstatistical samples may be similar, other characteristics of the sampling plan such as sample selection methods may not be. Further adjustments to the nonstatistical sample plan, for example, an increase in the sample size or changes in the selection method, may be needed to provide comparable assurance from statistical and nonstatistical plans.

4.67 An auditor might find familiarity with sample sizes based on statistical theory helpful when applying professional judgment and experience in considering the effect of various planning considerations on sample size. The nonstatistical sampling approaches illustrated in this chapter are consistent with statistical sampling theory.

Examples of Sample Size Determination

4.68 Table 4-5, "Illustrative Sample Sizes," shows various sample sizes that might be used for statistical or nonstatistical sampling based on a MUS statistical approach.[21] The auditor using this table as an aid in understanding the relative size of samples for substantive tests of details needs to apply professional judgment in

- determining tolerable misstatement.
- estimating expected misstatement.
- quantifying the acceptable level of risk of incorrect acceptance.[22]
- estimating the population amount after the removal of items to be examined 100 percent.
- determining the appropriate sample size that would reflect differences between the nonstatistical approach and the MUS approach underlying the table and considering the aforementioned factors.

---

[21] Table 4-5, "Illustrative Sample Sizes," contains sample sizes for MUS given tolerable misstatement, expected misstatement, and the risk of incorrect acceptance. The table incorporates the conservative assumption that the total tainting consists of the maximum number of 100 percent tainted items plus, if necessary, 1 partially tainted item. For example, if risk is 5 percent, tolerable misstatement is 3 percent of the population, and expected misstatement is 40 percent of tolerable (in other words, 1.2 percent of the population), then the tabulated sample size is 270. This means that the expected sum of the taints is 3.24 (270 multiplied by 1.2 percent). Accordingly, the tabulated sample size is computed on the assumption that the sample will contain (3) 100 percent tainted items and (1) 24 percent tainted item. For a further discussion of taintings, see chapter 6 and table C-2, "Confidence Factors for Monetary Unit Sample Size Design," of appendix C.

[22] Paragraphs 4.39–.42 provide a discussion of the audit risk model.

Table 4-5
Illustrative Sample Sizes

Risk of Incorrect Acceptance	Ratio of Expected to Tolerable Misstatement	Tolerable Misstatement as a Percentage of Population										Expected Sum of Taints
		50%	30%	10%	8%	6%	5%	4%	3%	2%	1%	0.50%	
5%	—	6	10	30	38	50	60	75	100	150	300	600	—
5%	0.10	8	13	37	46	62	74	92	123	184	368	736	0.37
5%	0.20	10	16	47	58	78	93	116	155	232	463	925	0.93
5%	0.30	12	20	60	75	100	120	150	200	300	600	1,199	1.80
5%	0.40	17	27	81	102	135	162	203	270	405	809	1,618	3.24
5%	0.50	24	39	116	145	193	231	289	385	577	1,154	2,308	5.77
10%	—	5	8	24	29	39	47	58	77	116	231	461	—
10%	0.20	7	12	35	43	57	69	86	114	171	341	682	0.69
10%	0.30	9	15	44	55	73	87	109	145	217	433	866	1.30
10%	0.40	12	20	58	72	96	115	143	191	286	572	1,144	2.29
10%	0.50	16	27	80	100	134	160	200	267	400	799	1,597	4.00
15%	—	4	7	19	24	32	38	48	64	95	190	380	—
15%	0.20	6	10	28	35	46	55	69	91	137	273	545	0.55
15%	0.30	7	12	35	43	57	69	86	114	171	341	681	1.03
15%	0.40	9	15	45	56	74	89	111	148	221	442	883	1.77
15%	0.50	13	21	61	76	101	121	151	202	302	604	1,208	3.02
20%	—	4	6	17	21	27	33	41	54	81	161	322	—
20%	0.20	5	8	23	29	38	46	57	76	113	226	451	0.46
20%	0.30	6	10	28	35	47	56	70	93	139	277	554	0.84
20%	0.40	8	12	36	45	59	71	89	118	177	354	707	1.42
20%	0.50	10	16	48	60	80	95	119	159	238	475	949	2.38
25%	—	3	5	14	18	24	28	35	47	70	139	278	—
25%	0.20	4	7	19	24	32	38	48	64	95	190	380	0.38
25%	0.30	5	8	23	29	39	46	58	77	115	230	460	0.69
25%	0.40	6	10	29	37	49	58	73	97	145	289	578	1.16
25%	0.50	8	13	38	48	64	76	95	127	190	380	760	1.90
30%	—	3	5	13	16	21	25	31	41	61	121	241	—
30%	0.20	4	6	17	21	27	33	41	54	81	162	323	0.33
30%	0.40	5	8	24	30	40	48	60	80	120	239	477	0.96
30%	0.60	9	15	43	54	71	85	107	142	213	425	850	2.55
35%	—	3	4	11	14	18	21	27	35	53	105	210	—
35%	0.20	3	5	14	18	23	28	35	46	69	138	276	0.28
35%	0.40	4	7	20	25	34	40	50	67	100	199	397	0.80
35%	0.60	7	12	34	43	57	68	85	113	169	338	676	2.03
50%	—	2	3	7	9	12	14	18	24	35	70	139	—
50%	0.20	2	3	9	11	15	18	22	29	44	87	173	0.18
50%	0.40	3	4	12	15	19	23	29	38	57	114	228	0.46
50%	0.60	4	6	17	22	29	34	43	57	85	170	340	1.02

4.69 Table 4-5 might also help the auditor understand the risk level implied by a given sample size. For example, the auditor might be designing a nonstatistical sampling application to test a population of 2,000 accounts receivable balances with a total recorded amount of $1 million. The auditor may have

- considered selecting a sample of 60.
- determined tolerable misstatement to be $50,000 (5 percent of the population).
- expected no misstatements in the sample.

Table 4-5 indicates that the sample of 60 implies a 5 percent risk of incorrect acceptance if no misstatements are found.

4.70 The auditor might also compare other sample sizes in the table with the sample size of 60 to gain a better understanding of how sample size affects the risk levels in the circumstances. The auditor using table 4-5 for this purpose also applies professional judgment in assessing the factors described in the preceding paragraph.

4.71 The calculation of 60 sampling units is based on a stratified sampling (or MUS, using a PPS selection technique) approach. The sample size would be appropriate if the auditor uses such an approach in selecting the sample. If selecting the sample on an item (not dollar) basis, stratification may be particularly important to increasing the efficiency of the sample. If the nonstatistical sample design is planned without stratification (or PPS selection), the auditor increases the sample size. For example, in the absence of stratification, the sample of 60 items might be increased to 90 items if consideration of the diversity of values in the population leads the auditor to conclude a 50 percent increase is appropriate.

4.72 A simple formula approach can also be used to determine a nonstatistical sample size. The simple formula is comprised of three elements—the population's recorded amount, a confidence factor (assurance factor), and tolerable misstatement. Factors for other risk levels are noted in the zero expected misstatement line in table C-2, "Confidence Factors for Monetary Unit Sample Size Design," of appendix C.

$$\text{Sample Size} = \frac{\text{Population Recorded Amount} \times \text{Confidence Factor}}{\text{Tolerable Misstatement}}$$

Table 4-6

Confidence (Reliability) Factors

Risk of Incorrect Acceptance (%)	Confidence of Sample (%)	Confidence Factor
37	63%	1
14	86%	2
5	95%	3

For purposes of the following illustration, expected misstatement is expected to be zero and the population is assumed to be large.

4.73 As an example, suppose the auditor using the formula approach has a population of $100,000 and a tolerable misstatement of $3,000, expected misstatement is zero, and an acceptable risk of incorrect acceptance of 14 percent for an assurance factor of 2. The sample size using the formula is 67 items (67 = [$100,000 = 2] ÷ $3,000).

4.74 The formula produces samples sizes identical to table 4-5 when expected misstatement is zero. When the auditor expects some misstatement, various approaches may be used to adjust the sample size.[23] Some auditors use table 4-5 when they expect misstatements. Others use informed judgment or a rule-of-thumb to adjust the sample size for some expected misstatement. Other auditors calculate a more precise sample size by using the additional confidence factors (in other words, assurance factors or reliability factors) provided in table C-1, "Monetary Unit Sample Size Determination Tables," and table C-2 of appendix C or by using the formula approach illustrated in chapter 6 for MUS samples or the formula approach described and illustrated in table C-4, "Alternative MUS Sample Size Determination Using Expansion Factors," of appendix C. Any of these methods, properly applied, can result in adequate sample sizes. For identical risks of incorrect acceptance, sample sizes determined by table 4-5 (table C-1 in appendix C) and table C-2 in appendix C will be the same.

Performing the Sampling Plan

4.75 The auditor should perform auditing procedures that are appropriate for the particular audit objectives to each sample item. In some circumstances, the auditor might not be able to apply the planned procedures to selected sampling units (for example, because the client could not locate the supporting documentation). The auditor's treatment of those unexamined items depends on their effect on the evaluation of the sample. If the auditor's evaluation of the sample results would not be altered by considering those unexamined items to be misstated, it is not necessary to examine the items; however, if considering those unexamined items to be misstated would lead to a preliminary conclusion that the balance or class of transactions is materially misstated, the auditor should consider alternative procedures that would provide sufficient evidence to form a revised conclusion. The auditor also should consider whether the reasons for the inability to examine the items affect the planned assessed *risks of material misstatement* or the auditor's assessment of the risk of fraud.

4.76 Some of the selected sampling units might be unused or voided items. The auditor should consider how the population has been defined when he or she decides whether to include such an item in the sample. If the population consists of all checks, whether issued or voided, the auditor may need to consider the possibility that the sample of checks will contain one or more voided checks. If the auditor excludes these voided items from the sample evaluation,[24] then the number of valid sample units selected will be less than what was desired. To provide for this possibility, the auditor might wish to select a slightly larger number of sample items. The additional items would be examined only if they were used as replacement items.

---

[23] As expected misstatement increases, this formula will result in sample sizes that will likely return lower confidence levels than desired. If the auditor desires to maintain the planned level of confidence, then the auditor may need to increase the sample size.

[24] For example, when the voided items would not contain the characteristic of interest such as a recorded amount: a sample of 20 checks with 2 voided items would be evaluated as a sample of 18.

Evaluating the Sample Results

Projecting the Misstatement to the Population

4.77 Paragraph .13 of AU-C section 530 states that "the auditor should project the results of audit sampling to the population." To that amount the auditor should add any misstatements discovered in any items examined 100 percent.

4.78 Regardless of whether the sample results support the assertion that the recorded amount is not misstated by an amount greater than tolerable misstatement, the auditor should request management to record the factual misstatements identified in the population unless clearly trivial;[25] however, even if the entity does correct all factual misstatements, that does not eliminate the need to consider the remaining projected misstatement.

4.79 In determining the adequacy of evidence from the sample, the total factual and projected misstatement,[26] adjusted for misstatements corrected by the entity, may be compared with the tolerable misstatement for the account balance or class of transactions. If the total factual and projected misstatement is less than the tolerable misstatement for the account balance or class of transactions, the auditor then should consider the risk that such a result might be obtained even though the true monetary misstatement for the population exceeds the tolerable misstatement. Alternatively, the auditor may compare the projected misstatement to the expected misstatement used in determining the sample size. When the projected misstatement exceeds the expected misstatement, the sample may not be large enough to provide the planned risk of incorrect acceptance (that is, results in an inadequate allowance for sampling risk.)In other words, pursuant to paragraph .A27 of AU-C section 530, the auditor typically considers the risk (for instance, sampling risk) that there might be other, undetected misstatements remaining in the population examined that might indicate a material misstatement or an amount greater than tolerable misstatement exists (also refer to the guidance in paragraph .A5 of AU-C section 450).

4.80 When nonstatistical methods are used, this consideration of sampling risk is made using informed judgment. The auditor, in making this judgment, would consider not only the results of procedures, but the nature, timing, and extent of procedures performed that led to the test result.

4.81 The auditor should also aggregate the projected misstatement in the balance or class (after adjustments, if any) with other factual and projected misstatements in other balances and classes to evaluate whether the financial statements as a whole may be materially misstated. Paragraph .11 of AU-C section 450, *Evaluation of Misstatements Identified During the Audit* (AICPA, *Professional Standards*), establishes requirements and provides guidance for the auditor when evaluating the effect of uncorrected misstatements.

---

[25] Pursuant to paragraph .12 of AU-C section 450, *Evaluation of Misstatements Identified During the Audit* (AICPA, *Professional Standards*), audit documentation should include the amount the auditor has deemed to be trivial.

[26] The sum of total factual misstatement and total projected misstatement is the difference between the estimated amount of the account balance or class of transactions being examined and the entity's recorded amount. Factual misstatement is specifically identified misstatement, such as a difference identified in a sample item or an item examined 100 percent. Projected misstatement is generally developed by extrapolation from the factual misstatements in sample items.

4.82 There are several methods the auditor can use to project the amount of misstatement found in a statistical or nonstatistical sample to estimate the amount of misstatement in the population. When choosing the method of projection, the auditor may consider the method of sample selection. For example, a sample designed and selected using MUS sampling concepts (whether statistical or nonstatistical) would suggest that a MUS methodology be used to project the sample results. Similarly, a stratified item-based sample would suggest the use of a comparable sample projection methodology (for example, difference or ratio projection). When statistical sampling is used, a statistically valid sample evaluation approach appropriate to the sampling approach applied is followed. When nonstatistical methods are used, similar approaches may be applied. This section describes three potential projection methods.[27]

4.83 One method of projecting the amount of misstatement is to apply the misstatement rate of dollar misstatements observed in the sample to the population. For example, an auditor might have selected a sample that sums to $10,000 and observed an overstatement misstatement of $100, or 1 percent of the recorded amount of the accounts-receivable balance tested. If the total recorded amount in the population is $100,000, then projected misstatement is $1,000 ($100,000 × 1 percent). The projection method based on the misstatement rate observed in the sampling population does not require an estimate of the number of sampling units in the population. If the auditor designed the sample by separating the items subject to sampling into groups or strata, he or she would project the misstatement results of each group separately and then calculate an estimate of misstatement in the population by summing the individually projected amounts from each group. The auditor would also add to the projected amount of misstatement any misstatement found in the high risk items that were examined 100 percent. This approach may be appropriate for most item-based[28] samples, if it is applied by strata. Where the sample approximates an informal MUS methodology (without defined strata), the auditor may apply the MUS method in paragraph 4.85.

4.84 Another method used to project misstatement to the population projects the average difference between the audited and the recorded amounts of each item in the sample to all items constituting the population. For example, the auditor might have selected a sample of 100 items. If the auditor found $200 of misstatement in the sample, the average difference between the audited and recorded amounts for items in the sample is $2 ($200 ÷ 100). The auditor then estimates the amount of misstatement in the population by multiplying the total number of items in the population (in this case, 5,000 items) by the average difference of $2 for each sample item. The auditor's estimate of the misstatement in this population is $10,000 (5,000 × $2). If the auditor designed the sample by separating the items subject to sampling into groups or strata, he or she should project the misstatement results of each group separately and then calculate an estimate of misstatement in the population by summing the individually projected amounts from each group. The auditor should also add to the projected amount of misstatement any misstatement found in the individually significant items that were examined 100 percent. This method may be

---

[27] Other methods may be appropriately used, but are beyond the scope of this guide. For example, another method of projection is based on projecting the audited sample amounts to result in a projected population total that is then compared to the recorded amount total. Another classical sampling projection method is based on the average per-sample-item ratio of audited to recorded sample item values, then projected to the total recorded value of the population.

[28] This approach approximates the ratio projection approach for classical variables samples.

appropriate for many item-based samples[29] if it is applied by strata. Where the sample approximates an informal MUS methodology (without defined strata), the auditor may apply the MUS method in paragraph 4.85.

4.85 When the nonstatistical sample selection has approximated a PPS selection, the auditor may also consider developing a point estimate drawing on the MUS method described in chapter 6. The point estimate could be obtained by first estimating a sampling interval by dividing the population dollars by the sample size. This interval would then be multiplied by each of the taintings obtained for any sample misstatements. The resulting products would be summed to obtain the PPS point estimate for the nonstatistical sample. The tainting for each misstatement is obtained by dividing each misstatement amount by its book value. Using the example from paragraphs 4.83–.84, the implicit sampling interval would be the book value of $100,000 divided by the sample size of 100 or an interval of $1,000. Two overstatements were found in balances of $100 (overstated by $100, a tainting of 1.0) and $200 (overstated by $100, a tainting of 0.5). In this case, the projected misstatement would be $1,500 [($1,000 × 0.5) + ($1,000 × 1.0)].

4.86 The auditor may choose between the projection approaches described in paragraphs 4.83–.84 on the basis of his or her understanding of the magnitude and distribution (pattern) of misstatements in the population. For example, if the auditor finds that the amount of misstatement relates closely to the size of an item, he or she ordinarily would choose the first approach (ratio). On the other hand, if the auditor finds the misstatements to be relatively constant for all items in the population, he or she might choose the second approach (difference). The various methods described will often give similar, but rarely identical, results when applied to the same sample result. If the difference between the results of the various methods is significant,[30] then the auditor may consider the possible reasons for the difference, such as considering the nature and size of the misstatements identified relative to the recorded amounts of the items for which the misstatements are identified. If the reasons for the difference can be discerned from the sample analysis, then that analysis may suggest the most appropriate technique for the projection. The assistance of sampling specialists can be helpful when it is not clear how to project the sample, or when both significant[31] understatements and overstatements are found in an MUS based sample.

Qualitative Factors

4.87 Paragraph .A23 of AU-C section 530 states the following:

> In addition to the evaluation of the frequency and amounts of monetary misstatements, [AU-C] section 450 requires the auditor to consider the qualitative aspects of the misstatements These include (*a*) the nature and cause of misstatements, such as whether they are differences in principle or in application, are errors or are caused by fraud, or are due to misunderstanding of instructions or carelessness, and (*b*) the possible relationship of the misstatements to other phases of the audit. ...

---

[29] This approach approximates the difference projection approach for classical variables samples.

[30] There is no requirement to compute the result under the various methods and compare them.

[31] Significant in amount or relative size (for example, a greater than 100 percent misstatement) to the item examined.

Nonstatistical and Statistical Audit Sampling for Substantive Tests of Details

4.88 A significant list of factors is cited in paragraph .A23 of AU-C section 450 that might cause the auditor to evaluate a misstatement as material. Circumstances that may affect the evaluation include the extent to which the misstatement

- affects compliance with regulatory requirements.
- affects compliance with debt covenants or other contractual requirements.
- relates to the incorrect selection or application of an accounting policy that has an immaterial effect on the current period's financial statements but is likely to have a material effect on future periods' financial statements.
- masks a change in earnings or other trends, especially in the context of general economic and industry conditions.
- affects ratios used to evaluate the entity's financial position, results of operations, or cash flows.
- affects segment information presented in the financial statements (for example, the significance of the matter to a segment or other portion of the entity's business that has been identified as playing a significant role in the entity's operations or profitability).
- has the effect of increasing management compensation (for example, by ensuring that the requirements for the award of bonuses or other incentives are satisfied).
- is significant with regard to the auditor's understanding of known previous communications to users (for example, regarding forecast earnings).
- relates to items involving particular parties (for example, whether external parties to the transaction are related to members of the entity's management).
- is an omission of information not specifically required by the applicable financial reporting framework but that, in the professional judgment of the auditor, is important to the users' understanding of the financial position, financial performance, or cash flows of the entity.
- affects other information that will be communicated in documents containing the audited financial statements (for example, information to be included in a "Management Discussion and Analysis" or an "Operating and Financial Review") that may reasonably be expected to influence the economic decisions of the users of the financial statements. [AU-C] section 720, *Other Information in Documents Containing Audited Financial Statements* [(AICPA, *Professional Standards*),] addresses the auditor's consideration of other information, on which the auditor has no obligation to report, in documents containing audited financial statements.
- is a misclassification between certain account balances affecting items disclosed separately in the financial statements (for example, misclassification between operating and nonoperating income or recurring and nonrecurring income items or a misclassification between restricted and unrestricted resources in a not-for-profit entity).

- offsets effects of individually significant but different misstatements.
- is currently immaterial and likely to have a material effect in future periods because of a cumulative effect, for example, that builds over several periods.
- is too costly to correct. It may not be cost beneficial for the client to develop a system to calculate a basis to record the effect of an immaterial misstatement. On the other hand, if management appears to have developed a system to calculate an amount that represents an immaterial misstatement, it may reflect a motivation of management.
- represents a risk that possible additional undetected misstatements would affect the auditor's evaluation.
- changes a loss into income or vice versa.
- heightens the sensitivity of the circumstances surrounding the misstatement (for example, the implications of misstatements involving fraud and possible instances of noncompliance with laws or regulations, violations of contractual provisions, and conflicts of interest).
- has a significant effect relative to reasonable user needs (for example,
    — earnings to investors and the equity amounts to creditors,
    — the magnifying effects of a misstatement on the calculation of purchase price in a transfer of interests [buy-sell agreement], and
    — the effect of misstatements of earnings when contrasted with expectations).
- relates to the definitive character of the misstatement (for example, the precision of an error that is objectively determinable as contrasted with a misstatement that unavoidably involves a degree of subjectivity through estimation, allocation, or uncertainty).
- indicates the motivation of management (for example, [i] an indication of a possible pattern of bias by management when developing and accumulating accounting estimates, [ii] a misstatement precipitated by management's continued unwillingness to correct weaknesses in the financial reporting process, or [iii] an intentional decision not to follow the applicable financial reporting framework).

4.89 As stated in paragraph .A22 of AU-C section 450, "determining whether a classification misstatement is material involves the evaluation of qualitative considerations, such as the effect of the classification misstatement on debt or other contractual covenants, the effect on individual line items or subtotals, or the effect on key ratios." Thus, in some circumstances a misclassification (only affects the balance sheet and not the income statement) may be larger than the materiality amount set for the audit and still have no material effect on the overall fairness of the financial statements.

The Sufficiency of Sampling Evidence for Proposing Adjustments

4.90 When considering the sufficiency of evidence supporting a projection or a proposed adjustment, the auditor may consider the extent of testing underlying the projected misstatement and the resultant ability of the sample to provide precise results. Paragraph .A24 of AU-C section 530 states: "Due to sampling risk, this projection may not be sufficient to determine an amount to be recorded." For example, any sample result can be projected to a population, however small the sample. But small samples may lack precision in estimating the audited amount. The client or the auditor might consider additional evidence to be necessary to support a projected material misstatement or proposed adjustment if the sample size supporting the projection was small (for example, less than 20 items). An auditor using statistical methods obtains a numerical precision or range that indicates how close the point estimate from the sample might be to the true population parameter (for example, the true amount of misstatements in the population). An auditor using nonstatistical methods uses judgment to estimate the precision of the projection. It is important to recognize that projections based on smaller sample sizes are likely to be imprecise.

Negative Confirmations

4.91 Because unreturned negative confirmations do not provide evidence that the intended third party received the request and verified that the information contained on it is correct, they rarely provide an adequate basis for projecting misstatement to the population of accounts. Paragraphs .A33–.A34 of AU-C section 505, *External Confirmations* (AICPA, *Professional Standards*), provide guidance for the auditor when evaluating the results from negative confirmations.

Interim Sample Results

4.92 A practical question that arises is whether interim sampling results can be projected or extrapolated from the interim population to that at year-end when the balance may not be the same. A sample should only be projected to the population from which it was selected. The auditor considers this question when determining any necessary further procedures. Accounts such as inventories and receivables change quite rapidly over time. Some fixed asset accounts, on the other hand, may not change much or at all between interim and year-end. In considering the evidence obtained from an interim audit sampling procedure and additional evidence that might be required, the auditor may also consider other factors in AU-C section 330, *Performing Audit Procedures in Response to Assessed Risks and Evaluating the Audit Evidence Obtained* (AICPA, *Professional Standards*).

Considering Sampling Risk at the Test Level

4.93 Paragraph .14 of AU-C section 530 states that "the auditor should evaluate the results of the sample, including sampling risk, and whether the use of audit sampling has provided a reasonable basis for conclusions about a population that has been tested." Paragraph .A27 of AU-C section 530 states the following:

> In the case of tests of details, the *projected misstatement* is the auditor's best estimate of misstatement in the population. As the projected misstatement approaches or exceeds tolerable misstatement,

the more likely that actual misstatement in the population exceeds tolerable misstatement. Also, if the projected misstatement is greater than the auditor's expectations of misstatement used to determine the sample size, the auditor may conclude that there is an unacceptable sampling risk that the actual misstatement in the population exceeds the tolerable misstatement.

Thus the total factual and projected misstatement for a sample is compared with the tolerable misstatement for the account balance or class of transactions, and appropriate consideration should be given to sampling risk. If the total projected misstatement is less than tolerable misstatement for the account balance or class of transactions, the auditor still should consider the risk that such a result might be obtained even though the true monetary misstatement for the population exceeds tolerable misstatement. For example, if the tolerable misstatement in an account balance of $1 million is $50,000 and the total projected misstatement based on an appropriately sized sample is $10,000, the auditor may be reasonably assured that there is an acceptably low sampling risk that the true monetary misstatement for the population exceeds tolerable misstatement. On the other hand, if the total projected misstatement is close to or exceeds the tolerable misstatement, the auditor may conclude that there is an unacceptably high risk that the actual misstatements in the population exceed the tolerable misstatement.

4.94 The auditor using nonstatistical sampling uses his or her experience and professional judgment in making such an evaluation; however, when the projected misstatement is close to tolerable misstatement, the auditor may typically conclude that there is an unacceptable risk that the true misstatement exceeds tolerable misstatement. When the projected misstatement identified in the audit test exceeds the auditor's expectation of the amount of misstatement used when designing the audit procedures, the auditor may typically conclude that there is a higher risk (than the level used in designing the test) that the true misstatement exceeds the tolerable misstatement.

4.95 The auditor may encounter an unexpected amount of projected misstatement compared to what was expected to be in the population. The auditor should investigate the nature and causes of the misstatements. In such cases, it is important for the auditor to recognize that the sample is expected to be representative only with respect to the incidence of misstatement in the population. Even if the misstatement appears to be from an unusual source, that does not mean that other unusual items are not in the population and that the original sample was not representative. The auditor might typically first evaluate the reasons for the misstatements and then assess whether, if the sample were extended, the evaluation for the combined samples would likely be sufficiently precise to support a conclusion with the desired level confidence. When sample results are extended, the original sample items and results typically are not discarded, but the additional sampling unit results are added to the original sample. Paragraphs 4.99–.102 provide additional discussion.

4.96 Extending the sample when the initial sample result was indicative of the true misstatement in the population will likely result in further misstatements being identified. If there is evidence that the misstatement was intentional or could be an indicator of a fraud, then the auditor would often carefully consider the appropriate next steps.

4.97 When seeking additional sampling evidence concerning the population, a rule of thumb used by some auditors is to at least double the original sample size to have much of an effect on the projected results or the allowance for sampling risk of the original sample. When the auditor uses statistical sampling, a more precise calculation of the needed sample expansion can be made.

4.98 If the sample results do not support the recorded amount of the population and the auditor believes the recorded amount might be misstated, the auditor would typically consider the misstatement along with other audit evidence in evaluating whether the financial statements may be materially misstated. The auditor may request that management examine the class of transactions, account balance, or disclosure to identify and correct the misstatements in the population.[32]

4.99 Paragraph .11 of AU-C section 450 establishes requirements for the auditor regarding the aggregation and assessment of misstatements.[33] When forming an audit opinion, paragraph .14 of AU-C section 700, *Forming an Opinion and Reporting on Financial Statements* (AICPA, *Professional Standards*) further states the following:

> In order to form that opinion, the auditor should conclude whether the auditor has obtained reasonable assurance about whether the financial statements as a whole are free from material misstatement, whether due to fraud or error. That conclusion should take into account the following:
>
> - ...
> - The auditor's conclusion, in accordance with [AU-C] section 450, *Evaluation of Misstatements Identified During the Audit*, about whether uncorrected misstatements are material, individually or in aggregate.
> - ...

4.100 If the sample results suggest that the auditor's sampling planning assumptions were in error, appropriate action should be taken.[34] For example, if the amount or frequency of misstatements discovered in a substantive test of details is greater than that expected based on the assessed level of control risk, the auditor considers whether the assessed level of control risk and the *risks of material misstatement* is still appropriate. For example, a large number of misstatements discovered in the confirmation of receivables might indicate the need to reconsider the assessed level of control risk related to receivables, sales, cash receipts, or credit memos. Depending on the reason for the higher than expected number of misstatements, the auditor may also decide to modify the audit tests of other accounts that were designed with control risk assessed at less than high. The auditor relates the evaluation of the sample to other relevant audit evidence when forming a conclusion about the related account balance or class of transactions.

---

[32] Paragraph .A9 of AU-C section 450.

[33] Misstatements include factual, projected and judgmental misstatements, as defined by paragraph .A3 of AU-C section 450.

[34] In accordance with paragraphs .24 and .A73 of AU-C section 330, the auditor should reevaluate the strategy at any point in the audit process when evidence indicates the planning assumptions may have been in error.

Misstatements Not Projected[35]

4.101 Paragraph .13 of AU-C section 530 states that "the auditor should project the results of audit sampling to the population." Paragraph .A4 of AU-C section 450 further states that "a misstatement may not be an isolated occurrence. Evidence that other misstatements may exist include, for example, when the auditor identifies that a misstatement arose from a breakdown in internal control or from inappropriate assumptions or valuation methods that have been widely applied by the entity."

4.102 When an auditor selects a sample believed to be representative, it is appropriate to project all misstatements in the sample to the population. A misstatement due to "a clerical error" would rarely qualify for special treatment because it represents a general condition that was identified in a representative sample. Such errors may occur at other times in the year and under other circumstances. Even when the specific misstatements found in the population might be associated with a specific reason (for example, "the bookkeeper was on vacation"), similar misstatements not found in the sample may still exist in the population. A representative sample is generally expected to be representative of the population only with respect to the incidence of misstatements, and not necessarily with respect to their nature. Thus, the projection of all sample misstatements is supported.

4.103 In some limited instances it may be acceptable to not extrapolate (project) a misstatement from a sample to the whole population. For example, this may be because the population was improperly specified at the outset, and the auditor did not recognize the separate character of some of the transactions. For example, suppose a sample misstatement of a revenue item appears to be solely due to a faulty procedure for accounting for foreign currency transactions. However, only a few transactions of the entity ever involve foreign currency, and these are in the aggregate insignificant. Had the auditor carefully considered the population when setting up the test, foreign currency transactions might have been examined separately for this issue. In such a case, it may be acceptable not to project the foreign currency issue misstatement to the entire population of revenue but to relate the foreign currency error from the sample to the foreign currency transactions from the population. The auditor may then consider the sufficiency of evidence concerning this issue, and whether additional evidence might also be needed regarding other possible misstatements that might be in the sub-population of foreign currency transactions or in the general population.[36] Based on this assessment, the auditor may consider the sufficiency of evidence concerning these issues, and may decide to perform additional procedures to determine more precisely the nature and extent of misstatement due to the foreign currency procedure or any

---

[35] Various terms are found that relate to this concept such as *isolated instance*, *carve-out*, and *containment*.

[36] Paragraph .A22 of AU-C section 530 states the following:

> In analyzing the deviations and misstatements identified, the auditor may observe that many have a common feature (for example, type of transaction, location, product line, or period of time). In such circumstances, the auditor may decide to identify all the items in the population that possess a common feature and extend audit procedures to these items. In addition, such deviations or misstatements may be intentional and may indicate the possibility of fraud.

other additional misstatement conditions. When there is sufficient representation of such transactions in the original sample[37] or in an expanded sample of these transactions, the identified currency misstatements in a sample may be projected to the sub-population of foreign currency transactions. The original (whole) population would typically still be used for projecting any other types of misstatements not due to the foreign currency or related unique sub-population issues, such as a failure to properly account for revenue recognition under generally accepted accounting principles. In the aggregation of sample results, the foreign currency factual and projected misstatement would be aggregated with the projected misstatements from other types of misstatements.

4.104 Careful explanation should be made of the reasoning supporting not projecting a misstatement to the population. Evidence should clearly support this different treatment. The auditor is cautioned that such circumstances are likely to be limited. Care needs to be taken not to "explain away" identified misstatements. Note that a sample is usually expected to be representative only with respect to the occurrence rate or incidence of misstatements, not their specific nature (see also paragraph 1.05 in chapter 1, "Characteristics of Audit Sampling" and appendix G, "Glossary"). Just because a reason can be stated for the misstatement, does not mean it is justified to assume that that misstatement can be addressed and the population is thereby corrected. Other misstatement types may be in the population and not in the sample. There are risks to the client, users of financial statements, and to the auditor when misstatements are incorrectly handled.

Documenting the Sampling Procedure

4.105 AU-C section 230, *Audit Documentation* (AICPA, *Professional Standards*), provides general guidance on the documentation of audit procedures. Although AU-C section 450 identifies three specific documentation requirements, AU-C section 530 does not identify specific documentation requirements for samples. However, examples of items that the auditor may document for substantive audit samples are found in paragraph .12 of AU-C section 450 and listed in the following paragraph.

4.106 According to paragraph .12 of AU-C section 450, the auditor should include in the audit documentation

    a. the amount below which misstatements would be regarded as clearly trivial;

    b. all misstatements accumulated during the audit and whether they have been corrected; and

    c. the auditor's conclusion about whether uncorrected misstatements are material, individually or in the aggregate, and the basis for that conclusion.

4.107 The following are commonly documented for audit samples:

- The objectives of the test and the accounts and assertions affected
- The definition of the population and the sampling unit, including how the auditor considered the completeness of the population

---

[37] When a significant number of sample items relate to an insignificant portion of the population, the auditor may reconsider whether the sample is "representative" of the population.

- The definition of a misstatement
- The risk of incorrect acceptance or level of desired assurance (confidence)
- The risk of incorrect rejection, if used
- Estimated and tolerable misstatement
- The audit sampling technique used
- The method used to determine sample size
- The method of sample selection
- Identification of the items selected
- A description of how the sampling procedure was performed and a list of misstatements identified in the sample
- The evaluation of the sample (for example, projection and consideration of sampling risk)
- A summary of the overall sample conclusion (if not evident from the results)
- Any qualitative factors considered significant in making the sampling assessments and judgments

4.108 Paragraph .A14 of AU-C section 230 provides several examples regarding how an auditor can identify selected sample items in audit documentation.

Chapter 5

Nonstatistical Sampling Case Study

> **⊚ Update 5-1 *Audit*: Clarified Auditing Standards**
>
> The auditing guidance in this guide edition has been conformed to Statement on Auditing Standards (SAS) Nos. 122–125, which were issued in 2011 as part of the Auditing Standards Board's Clarity Project. These clarified SASs are effective for audits of financial statements for periods ending on or after December 15, 2012. Although extensive, the revisions to generally accepted auditing standards resulting from these clarified SASs do not change many of the requirements found in the auditing standards that they supersede.
>
> To assist auditors and financial reporting professionals in making the transition, this guide includes appendix F, "Mapping and Summarization of Changes—Clarified Auditing Standards," which provides a cross reference of the sections in the superseded auditing standards to the applicable sections in the clarified auditing standards and identifies the changes, either substantive or primarily clarifying in nature, that may affect an auditor's practice or methodology relative to the applicable sections of SAS Nos. 122–125. It also summarizes the changes resulting from the requirements of SAS Nos. 122–125.
>
> The preface of this guide and the Financial Reporting Center on www.aicpa.org provide more information on the Clarity Project. Visit www.aicpa.org/sasclarity.

5.01 This chapter provides a case study illustrating the design and use of a nonstatistical sample.

5.02 Sarah Jones of Jones & Co., CPAs, designed a nonstatistical sample to test the existence and gross valuation of the December 31, 20XX, accounts-receivable balance of Short Circuit Inc., a privately owned electrical supply company that is a continuing client of Jones & Co. For the year ended December 31, 20XX, Short Circuit had sales of approximately $25 million. As of December 31, there were 905 accounts receivable, with debit balances aggregating $4.25 million. These balances ranged from $10 to $140,000. There were also 40 credit balances aggregating $5,000.[1]

5.03 In planning her audit, Sarah Jones updated her understanding of the client and its environment, including its internal control. She also understood that the entity's revenue recognition policy was to recognize revenue upon shipment. She also understood that cash sales are prohibited, and the entry for all sales transactions involves a debit to accounts receivable. In addition, all cash receipts are through the bank's lock box, and there are no credits to income in the cash receipts journal. The only general journal entries affecting receivables and revenue involve minor write-offs of bad debts and setting up an allowance for doubtful accounts at the end of each quarter. All of the preceding were true in prior audits, and inquiry of client management indicates no changes from prior periods.

---

[1] The net population consisted of 945 balances with a total recorded value of $4,245,000.

5.04 Jones made the following judgments in planning her procedures for revenue and receivables:

- Because this is not a first audit and because of some past errors in accounts receivables, her assessment of the *risks of material misstatement* in receivables did not support an assessment much below high for the assertions of existence and gross valuation of accounts receivable.
- Fraud risk related to revenue and receivables is low. There is little incentive to misstate revenue or receivables. The lock box system significantly reduces the risk of misappropriation of cash. The company's revenue recognition policy was appropriate in the circumstances. There was minimal risk of channel stuffing or other revenue recognition issues.
- The confirmation process would test existence and gross valuation of receivables. It would not provide much evidence about completeness, net valuation of receivables, presentation, and disclosure. Other procedures will be performed on the account that will address these assertions and provide some evidence on existence and valuation.
- The confirmation process would provide some evidence of the occurrence and gross valuation of sales transactions. This was because if receivables did not exist, the sales transaction did not occur. It also would provide some evidence about receivables cutoff because customers would report items included in receivables that were not shipped by year-end. The confirmation process would not provide evidence of completeness of revenue.

5.05 Sarah Jones made the following judgments in designing the confirmation sample:

- Her limited tests of controls supported an assessed level of *risk of material misstatement* (inherent and control risk) of high for the assertions of existence and gross valuation of accounts receivable.
- The preliminary assessment of overall materiality is $200,000.
- Tolerable misstatement for this test was set at $150,000, which is 75 percent of the materiality for the financial statements as a whole and less than the *performance materiality* set for the overall engagement. This judgment was based on the fact that most other accounts could be estimated to a significant precision and that the client would adjust for factual misstatements and follow up appropriately on projected misstatement issues, and this account had a higher expected error rate than other account balances.[2]
- She expects a possible $15,000 misstatement in accounts receivable, which is a realistic, or if anything, a somewhat conservative estimate based on the results of prior years' testing.
- The credit balances in accounts receivable would be tested separately.
- The balance for each selected customer would be confirmed.

---

[2] Had all of these relevant factors been considered when setting performance materiality, then performance materiality and tolerable misstatement may have been set to the same amount.

5.06 She planned to project the sample result using a ratio method. She made this judgment because she believed that the amount of misstatement in the population would be more likely to correlate to the total dollar amount of items in the stratum or population than to the number of items in the stratum or population.

5.07 The following is some additional information:

- The population contained 5 balances of more than $50,000, which totaled $500,000. Jones decided to examine these 5 balances 100 percent and exclude them from the population to be sampled. The population also contained 900 other debit balances, which totaled $3.75 million.

- Through substantive analytical procedures and cut-off tests, Jones obtained some assurance that all shipments were billed and that receivables were complete.

- The analytical procedures also provided some assurance for the assertions of existence and gross valuation of accounts receivable (for instance, the same assertions as the confirmation procedure).

Determining the Sample Size

5.08 Considering the following factors, Jones determined the sample size:

a. *Variation in the population.* Jones separated the population into 2 groups based on the recorded amounts of the items constituting the population. The first group consisted of 250 balances equal to or greater than $5,000 (total recorded amount of $2.5 million), and the second group consisted of the remaining balances that were less than $5,000 (total recorded amount of $1.25 million).[3] A computer program designed to interrogate populations electronically and select samples was used to efficiently perform this procedure.

b. *Risk that a material misstatement does not exist, when it does.* Referring to table 4-5, "Illustrative Sample Sizes," Jones decided to use a 5 percent *risk of incorrect acceptance* (that is, desired confidence level of 95 percent). She based her decision on an assessed level of *risk of material misstatement* of high, limited assurance from substantive analytical procedures, and because she did not plan any other significant detailed substantive or control tests to achieve the same objectives.[4]

c. *Tolerable and expected misstatement.* As indicated previously, the amount of tolerable misstatement for this sample was determined to be $150,000 and the amount of expected misstatement is estimated to be $15,000.

---

[3] Had the population not been *stratified*, the sample size would have been increased (see chapter 4, "Nonstatistical and Statistical Audit Sampling for Substantive Tests of Details") due to the variability of the items in the population.

[4] Had control tests been performed and supported effective controls, an acceptable risk higher than 5 percent (lower desired assurance) would likely have been appropriate. The extent of reduced assurance for this substantive test would be responsive to the extent of controls testing and the control test results. The design and performance of effective analytical procedures, for example by meaningful subclasses of receivables, can also reduce the extent of substantive detailed testing.

5.09 Jones then determined the appropriate sample size of 92 by referring to table 4-5 (5 percent risk that a sample will lead the auditor to conclude that a misstatement does not exist, when it does [risk of incorrect acceptance], 4 percent tolerable misstatement as a percent of the population, and 10 percent expected misstatement as a percent of tolerable misstatement).

5.10 Jones considered the efficiency of other strategy alternatives to reduce the sample size and concluded the confirmation procedures would be the most efficient and effective approach. She considered

- performing further tests of controls;
- performing additional detailed, targeted analytical procedures;
- performing a test of sales transactions (a related financial statement area) that would also provide assurance on assertions relevant to this test; or
- increasing the number of items that are substantively tested 100 percent by lowering the threshold for selecting items for 100 percent testing below $50,000.

However, she concluded that it would be more efficient to confirm 92 items.

5.11 She also decided to allocate the sample between the 2 groups in a way that was approximately proportional to the recorded amounts of the accounts in the groups. Accordingly, Jones selected on a haphazard basis 62 of the 92 customer balances from the first group or stratum (balances with recorded amounts equal to or greater than $5,000) and the remaining 30 customer balances from the second group or stratum (balances with recorded amounts under $5,000).

Evaluating the Sample Results

5.12 Jones mailed confirmation requests to each of the 92 customers whose balances had been selected and to each of the 5 customers selected in the 100 percent examination group. Of the 97 confirmation requests, 82 were completed and returned to her. She was able to obtain sufficient assurance through alternative procedures that the 15 customer balances that were not confirmed were bona fide receivables and were not misstated. Of the 82 responses, 4 customers indicated that their balances were overstated.[5] Jones investigated these balances further and concluded that they were, indeed, partially misstated. She determined that the misstatements resulted from ordinary mistakes (for example, shipping charge variations, a misapplication of discount agreements, and credits) in the accounting process. The sample was summarized as shown in table 5-1, "100 Percent Examination and Sample Testing Summary."

---

[5] Three were in the sample group and one was in the 100 percent tested group.

Table 5-1
100 Percent Examination and Sample Testing Summary

Group	Recorded Amount	Recorded Amount of Items Tested	Audited Amount of Items Tested	Amount of Overstatement
100% examination	$500,000	$500,000	$499,000	$1,000
Over $5,000	2,500,000	739,000	738,700	300
Under $5,000	1,250,000	62,500	62,350	150
	$4,250,000	$1,301,500	$1,300,050	$1,450

5.13 Jones observed that the sample included 30 percent of the dollar amount of the over $5,000 group and 24 percent of the items included in that group. She also observed that the sample comprised 5 percent of the dollar amount of the under $5,000 group, and about 5 percent of the items included in that group. On the basis of the preceding computations, she considered the methods of projecting sample results described in this guide.[6] She considered the misstatements found and confirmed her previous judgment that the amount of misstatement in the population was more likely to correlate to the total dollar amount of items in the stratum or population than to the number of items in the stratum or population. Thus, Jones decided to project misstatements based on the rate of misstatement in each group (stratum). Then Jones separately projected the rate of misstatement found in each group's sample to the total dollars from that group. For the over $5,000 group, she projected the sample results for that group to the population by multiplying the misstatement rate observed in the sample by the recorded amount for that group. She calculated the projected misstatement to be approximately $1,015 ($2,500,000 × ($300 ÷ $739,000)). Similarly, Jones calculated a projected misstatement for the group under $5,000 to be approximately $3,000 ($1,250,000 × ($150 ÷ $62,500)). Therefore, the total factual and projected misstatement from the items tested 100 percent and items sampled was $5,015 ($1,000 + $1,015 + $3,000). Management of Short Circuit Inc. agreed to correct the factual misstatements of $1,450, resulting in a remaining projected misstatement of $3,565.

5.14 Jones compared the total factual and projected misstatement from the items tested 100 percent and items sampled of $5,015 with her $15,000 expectation of misstatement of accounts receivable and concluded that the sample results met the desired test objective. She also compared the remaining unadjusted misstatement ($3,565) with the $150,000 tolerable misstatement and determined that there was a small risk that this account could be misstated by more than the tolerable misstatement (of $150,000). In other words, there was an ample "cushion" between the tolerable misstatement and the remaining projected misstatement amounts to be able to conclude there is a *low risk of material misstatement* in the account. Jones investigated the nature and cause of the misstatements and determined that, as they resulted from

---

[6] Had Jones selected the sample attempting to approximate a probability proportional to size selection, the monetary unit sampling point estimator described in chapter 6, "Monetary Unit Sampling," might also be used.

explainable minor clerical error, they were not indicative of additional audit risk or a significant deficiency or material weakness in controls. AU-C section 265, *Communicating Internal Control Related Matters Identified in an Audit* (AICPA, *Professional Standards*), provides further guidance on evaluating the severity of control deficiencies identified in the audit.

5.15 Jones concluded that the sample results supported the recorded amount of the accounts-receivable balance; however, she did aggregate the remaining projected misstatement from the sample results with other factual and projected misstatements to evaluate whether the financial statements as a whole might have been materially misstated. Her evaluation of the potential material misstatement of the financial statements as a whole included considering qualitative factors, for example, trends and account relationships.

5.16 The items she examined 100 percent were not part of the sample. Therefore, any misstatements from these items represented factual misstatements. Because Short Circuit Inc. agreed to correct the $1,000 misstatement, there was no need to consider these items in evaluating whether the financial statements as a whole may have been materially misstated.

Chapter 6

Monetary Unit Sampling

> **© Update 6-1 *Audit*: Clarified Auditing Standards**
>
> The auditing guidance in this guide edition has been conformed to Statement on Auditing Standards (SAS) Nos. 122–125, which were issued in 2011 as part of the Auditing Standards Board's Clarity Project. These clarified SASs are effective for audits of financial statements for periods ending on or after December 15, 2012. Although extensive, the revisions to generally accepted auditing standards resulting from these clarified SASs do not change many of the requirements found in the auditing standards that they supersede.
>
> To assist auditors and financial reporting professionals in making the transition, this guide includes appendix F, "Mapping and Summarization of Changes—Clarified Auditing Standards," which provides a cross reference of the sections in the superseded auditing standards to the applicable sections in the clarified auditing standards and identifies the changes, either substantive or primarily clarifying in nature, that may affect an auditor's practice or methodology relative to the applicable sections of SAS Nos. 122–125. It also summarizes the changes resulting from the requirements of SAS Nos. 122–125.
>
> The preface of this guide and the Financial Reporting Center on www.aicpa.org provide more information on the Clarity Project. Visit www.aicpa.org/sasclarity.

6.01 Whereas chapter 4, "Nonstatistical and Statistical Audit Sampling for Substantive Tests of Details," provided the general setting for the use of sampling for substantive tests, this chapter focuses specifically on a statistical sampling approach called monetary unit sampling (MUS). MUS is a subset of a broader class of procedures sometimes referred to as probability proportional to size (PPS) sampling.[1] PPS samples share the characteristic of selecting sample items where the probability of an item's selection for the sample is proportional to its recorded amount. In this guide, the term *PPS* is used to describe a method of sample selection, whereas *MUS* is used to describe the sample size and evaluation methods discussed in this chapter.

6.02 As discussed in chapter 2, "The Audit Sampling Process," attributes sampling is typically used to reach a conclusion about a population in terms of a rate of occurrence. Variables sampling is typically used to reach conclusions about a population in terms of a dollar amount. MUS is a method that uses attributes sampling theory to express a conclusion in dollar amounts rather than as a rate of occurrence. Variations of MUS sampling are known as

---

[1] A classical variables probability proportional to size (PPS) sample may be evaluated based on classical sampling theory. It uses an assumption that enough (for example, 20–25) misstatements be found in the sample to support the normal distribution theory underlying this evaluation method. Often auditors plan to and indeed find few or no misstatements in samples, and thus the classical variables PPS method may not be appropriate in many audit situations; further discussion is beyond the scope of this guide. For further information, see Donald Roberts, *Statistical Auditing* (New York: AICPA, 1978): 116–119.

AAG-SAM 6.02

dollar-unit sampling, cumulative monetary amounts (CMA) sampling, and combined attributes/variables sampling.

6.03 MUS methods have been used in auditing since the early 1960s because they overcome some of the limitations of classical variables sampling techniques, such as the low misstatement rates of many accounting populations, and because of their simplicity compared to designing classical statistical techniques. They are often used for audit testing purposes in auditing.[2] For many estimation purposes (for example, to estimate a precise projection and confidence limits from sample information) or for engagements outside the usual audit context, where the sample will be the basis for a settlement in a dispute or will likely involve a discussion with parties that are nonauditors (such as when computing a damages estimate), careful consideration needs to be given on whether MUS or classical sampling techniques should be employed. Depending on the specific methodology followed by the auditor, many MUS approaches have been demonstrated by simulation studies to provide conservative results (in other words, they understate the true confidence level of the test or overstate the risk that a sample will lead the auditor to conclude that a misstatement does not exist, when it does).

Selecting a Statistical Approach

6.04 Both statistical approaches to sampling for substantive testing—classical variables sampling and MUS—can provide sufficient audit evidence to achieve the auditor's objective; however, in some circumstances, MUS may be more efficient than classical variables sampling.

Advantages

6.05 The advantages of MUS are as follows:

- MUS is generally easier to apply than classical variables sampling. Because MUS is based on attributes sampling theory, the auditor can easily calculate sample sizes and evaluate sample results manually or with the assistance of tables, as well as by using audit software. Sample selection can be performed with the assistance of either a computer program or a calculator.

- MUS does not require direct consideration of the population characteristics (for example, standard deviation of dollar amounts or normality of the population characteristics) to determine the appropriate sample size because the sample is selected based on each item having a chance of selection proportional to its size. The size of a MUS sample is not based on any measure of the estimated variation of audited amounts, because each monetary unit (for example, *dollar*) in the population is of the same size. The size of a classical variables sample is responsive to the variation, or standard deviation, of the characteristic of interest shared by the items in the total population (see the discussion in chapter 7, "Classical Variables Sampling").

---

[2] Because monetary unit sampling (MUS) was developed and adapted specifically for audit use, its application may be less familiar to some statisticians outside the audit community.

- MUS automatically selects a sample in proportion to an item's dollar amount; thus, stratification to reduce variability is unnecessary. The auditor using classical variables sampling usually needs to stratify the population to compute an efficient sample size.
- The MUS systematic sample selection described in this guide automatically identifies any item that is individually significant if its amount exceeds the sampling interval.
- If the auditor expects (and finds) no misstatements, MUS usually results in a highly efficient sample size.
- A MUS sample can be designed more easily and sample selection can begin before the final and full population is completely available.

6.06 Some of the circumstances in which MUS may be especially useful include the following:

- Accounts receivable confirmation (when unapplied credits are not significant in amount, quantity, or risk)
- Loans receivable confirmation (for example, real estate mortgage loans, commercial loans, and installment loans)
- Tests of investment security pricing compared to published prices
- Inventory price tests in which the auditor anticipates relatively few misstatements and the population is not expected to contain a significant number of large (relative to book amount) understatements
- Fixed-asset additions tests where existence is the primary risk

Disadvantages

6.07 The disadvantages of MUS sampling are as follows:

- MUS is not designed to test for the understatement of a population; and because the sample is selected "proportional to size," it is quite unlikely to select small recorded amounts and these amounts may be significantly understated. Of course, neither classical variables sampling nor MUS can select items that are not included in the population. With MUS, the approach to testing for the understatement of a population is to test a related (reciprocal)[3] population for overstatement; for example, the auditor might test disbursements made after the year end in order to test for the understatement of recorded accounts payable. When the expected understatements might be significant in number or large understatement taintings are expected, a classical variables approach may be more appropriate.
- The typical approach to MUS includes an assumption that the audited amount of a sampling unit is not less than zero or greater than the recorded amount. If the auditor anticipates understatements (even in a receivables confirmation application) or situations in which the audited amount will be less than zero, a MUS

---

[3] For example, sales recorded after year-end are considered to be a *reciprocal* population to sales recorded prior to year-end. Any recorded sales would be in either one or the other population.

approach may require special design considerations or may be inappropriate.

- If an auditor identifies understatements in a MUS sample, evaluation of the sample requires special considerations. Large understatements (for example, more than 100 percent of the recorded amount) may lead to projections that are invalid or inconclusive. In particular, it might not be appropriate to offset (net) understatements and overstatements.

- Selection of zero or negative balances requires special design considerations. For example, if the population to be sampled is accounts receivable, the auditor may need to segregate credit balances into a separate population for testing. If examination of zero balances is important to the auditor's objectives, he or she would need to test them separately using an item-based sampling technique because zero balances are not subject to MUS (PPS) selection.

- When misstatements are found, MUS evaluation may overstate the allowance for sampling risk at a given risk level. As a result, the auditor may be more likely to reject an acceptable recorded amount for the population.

- The auditor usually needs to cumulatively sum (add through) the population for the MUS (PPS) selection procedure illustrated in this guide; however, adding through the population usually will not require significant additional effort because the related accounting records are typically stored electronically and audit software to select samples is used. The auditor often needs to total the population anyway to determine whether it is complete and reconciles with the financial statements.

- As the expected amount of misstatement increases, the appropriate MUS sample size increases. In such circumstances the auditor may sometimes find classical variables techniques such as the difference or ratio technique more efficient.

- Many MUS methodologies are conservative in stating the confidence achieved and usually only compute one-sided upper bounds. Accordingly in considering the use of sampling techniques in circumstances outside the usual audit testing situation (for example, for estimating amounts), the auditor may find other sampling techniques more effective and efficient.

6.08 Some of the circumstances in which MUS sampling might not be the most effective or efficient approach include the following:

- Accounts receivable confirmation in which a large number of unapplied credits exist

- Inventory test counts and price tests for which the auditor anticipates a significant number of misstatements that can be both understatements and overstatements

- Conversion of inventory from first in, first out to last in, first out

- Populations where individual recorded amounts are not available

- Any application in which the primary objective is to estimate independently the amount of an account balance or class of transactions (note that independence issues may arise when auditor estimates are used in determining reported financial statement amounts)

Defining the Sampling Unit

6.09 MUS applies attributes sampling theory to reach dollar-amount conclusions by selecting sampling units proportional to their size. Essentially, MUS sampling gives each individual dollar in the population an equal chance of selection. This helps the auditor direct the audit effort toward larger balances or transactions. As a practical matter, however, the auditor does not examine an individual dollar within the population. For illustrative purposes, some auditors think of each dollar as a hook that snags the entire balance or transaction that contains it. The auditor following this methodology usually examines the entire balance or transaction that includes the selected dollar. The balance or transaction that the auditor examines is called a *logical unit*.

6.10 A MUS approach can also be used for performing tests of controls (for example, when performing a dual purpose test). MUS provides evidence in terms of the proportion of dollars being processed by the controls rather than the rates of deviation on an item basis. In a dual purpose test, the basis for the controls evaluation is the operation of the control and not just the substantive correctness of the recorded item. It is possible, for example, that a control failed to be applied to a transaction, but the control failure did not lead to a misstatement; thus, different controls and substantive conclusions can be reached on the same sample item.

Selecting the Sample

6.11 This section discusses systematic (for example, fixed-interval) selection with one random start.[4] This method is easy to apply when selecting a sample either manually or using computer software. Systematic selection involves dividing the population into equal groups of dollars, selecting a dollar from each group, and identifying the logical unit associated with the selected dollar from each group. Each group of dollars is a sampling interval.

6.12 AU-C section 530, *Audit Sampling* (AICPA, *Professional Standards*), suggests that the auditor examine separately those items of high risk or for which accepting some sampling risk is not justified. According to paragraph .A15 of AU-C section 530, "the auditor might first separately examine those items deemed to be of relatively high risk and then use audit sampling ... to form an estimate of some characteristic of the remaining population." For example, individually material items or high risk items might be selected and tested 100 percent before sampling the remainder. Sometimes this approach will actually reduce the overall extent of testing, particularly when the items removed are a relatively significant portion of the population, and the remaining items are used to plan the sample. If large items are not removed, and systematic (that is, fixed interval) sampling (for example, every nth dollar) is used to select the MUS sample, all items equal to or larger than the sampling interval will be automatically selected for testing.

---

[4] For a more complete discussion of other MUS and PPS selection and evaluation methods, see Donald Roberts, *Statistical Auditing* (New York: AICPA, 1978): 21–23.

6.13 To use the systematic selection method, the auditor selects a random number between one and the sampling interval, inclusive. This number is the *random start*. The auditor then begins adding the recorded amounts of the logical units throughout the population. The first logical unit selected is the one that contains the dollar amount corresponding to the random start. The auditor then selects each logical unit containing every Jth dollar thereafter (J represents the sampling interval). For example, if an auditor uses a sampling interval of $5,000, he or she selects a random number between $1 and $5,000, inclusive, such as the 2,000th dollar, as the random start. Then the 7,000th dollar ($2,000 + $5,000), then the 12,000th dollar ($2,000 + $5,000 + $5,000), and every succeeding Jth (in this case, 5,000th) dollar is selected until the entire population has been subject to sampling. The auditor therefore examines the logical units that contain the 2,000th, 7,000th, and 12,000th dollars and so on.

6.14 One drawback of fixed-interval selection is the risk that the interval could coincide with a pattern in the population. For example if in a weekly payroll of $200,000 where the population is the total payroll for the year, and the last five persons on the payroll register are supervisors, a sampling interval of around 200,000 might pick the same person, or all employees or all supervisors. Possible solutions to this risk include using multiple random starts, randomizing the population, or picking a random dollar from within each sampling interval.

6.15 Because every dollar has an equal chance of being selected, logical units having more dollars (that is, a larger recorded amount) have a greater chance of being selected. Conversely, smaller logical units have a smaller chance of being selected. All logical units with dollar amounts equal to or greater than the sampling interval are certain to be selected under the systematic selection method.

6.16 If the recorded amount of a logical unit is several times larger than the sampling interval, the logical unit might be selected more than once.[5] If that happens, the auditor will usually ignore the repeat selection and consider the logical unit only once when evaluating the sample results. Because logical units with recorded amounts greater than the sampling interval might be selected more than once, the actual number of logical units selected for the sample might be less than the computed sample size. That consideration is discussed further in this chapter. To address this issue and to refine the evaluation of errors found in the large items and the sampled items, auditors may also remove items that are equal to or larger than[6] the sampling interval for 100 percent examination.

6.17 Items in the population with negative balances require special consideration, usually because they have different risk characteristics. One way is to exclude them from the selection process and test them separately. Another approach is to change the sign of the negative items and add them to the positive population before selection, thereby *testing* the entire population

---

[5] There are various methods of PPS sample selection in use (such as the cell method and the random dollar selection method). With these methods, a logical unit (sample item) may be selected more than one time even if it is not larger than the sampling interval.

[6] Some auditors also remove items that are less than, but close to, tolerable misstatement as this sometimes reduces the total testing effort and protects against the risk that these larger accounts may contain misstatements that might aggregate to a material amount and might not be selected for examination.

in one sample. The latter approach is typically used only when there are few negative items and few or no misstatements expected, as the evaluation of misstatements involving negative items that were included in the population may necessitate the assistance of a statistical sampling specialist to interpret the results. Some auditors therefore use only the former approach.

6.18 If the selection is to be done manually (or with an electronic spreadsheet), the auditor can use a calculator or electronic spreadsheet in the following manner:

　　a. Clear the calculator.

　　b. Subtract the random start.

　　c. Begin adding the recorded amounts of logical units in the population, obtaining a subtotal after the addition of each succeeding logical unit. Items with negative balances are excluded. The first logical unit that makes the subtotal zero or positive is selected as part of the sample.

　　d. After each selection, subtract the sampling interval as many times as necessary to make the subtotal negative again.

　　e. Continue adding the logical units as before, selecting all items that cause the subtotal to equal zero or become positive.

6.19 The auditor may typically reconcile the total recorded amount of logical units accumulated on the calculator to a control total of the recorded amount of the population. The auditor then may add (*a*) the balance shown on the calculator, (*b*) the random start, and (*c*) the sampling interval multiplied by the number of times it was subtracted on the calculator. The total should be the control total for positive amounts. If it is not, either the population total is different from the control total or an error was made in selecting the sample. The auditor then usually corrects any errors in the sample selection.

Determining the Sample Size

6.20 One way that MUS sample sizes can be determined is by reference to table 4-5, "Illustrative Sample Sizes" (also, table C-1, "Monetary Unit Sample Size Determination Tables," in appendix C, "Monetary Unit Sampling Tables"). If the auditor chooses to use the table the auditor needs to determine an appropriate risk of incorrect acceptance and may need to express tolerable misstatement as a percentage of the population and expected misstatement (if any) as a percentage of tolerable misstatement.[7] For example, if a 90 percent confidence is desired (in other words, a 10 percent risk that a sample will lead the auditor to conclude that a misstatement does not exist, when it does (that is, risk of incorrect acceptance) and expected error (for example, 1 percent) is 20 percent of tolerable error (5 percent), then the resulting sample size is 69 items. Once the sample size has been determined, the sampling interval can be calculated by dividing the population size by the sample size. The sampling interval is often rounded down to a convenient number.

6.21 Table 4-5 also gives the sum of taintings that the auditor may find and still achieve the audit objectives. In this example, if the auditor uses a

---

[7] Some auditors use other methods to determine MUS sample sizes, such as other tables and computer programs. This guide does not discuss all the potential methods for determining MUS sample sizes.

sample size of 69 items, he or she may find total taintings of 0.69 and conclude at the desired risk (the complement of the confidence level) that the population was not misstated by more than tolerable misstatement.

6.22 Table 4-5 may be used when no misstatements are expected and when some misstatements are expected. As discussed in the following section, there are other methods for determining sample sizes.

Formula Method—No Misstatements Expected

6.23 The size of an appropriate sampling interval is related to the auditor's consideration of the risk that the sample will lead the auditor to conclude that material misstatement does not exist, when it does (that is, risk of incorrect acceptance) and the tolerable misstatement. If table 4-5 is not used to determine a sample size, some auditors calculate a sampling interval by dividing tolerable misstatement by a factor that corresponds to the risk of incorrect acceptance. The factor is known as the *confidence (reliability) factor*. Some such factors are presented in table 6-1, "Confidence (Reliability) Factors."

Table 6-1

Confidence (Reliability) Factors

Risk of Incorrect Acceptance (%)	Confidence of Sample (%)	Confidence Factor
37	63%	1
14	86%	2
5	95%	3

6.24 For example, if the auditor assesses the tolerable misstatement as $15,000, expected misstatement at zero, and the risk of incorrect acceptance as 5 percent, the sampling interval is calculated to be $5,000 ($15,000 ÷ 3). If the recorded amount of the population is $500,000, the sample size is 100 ($500,000 ÷ $5,000).

6.25 Table C-2, "Confidence Factors for Monetary Unit Sample Size Design," in appendix C provides factors for some commonly used risks of incorrect acceptance. The appropriate row to use with the guidance in this subsection, "No Misstatements Expected," is the row with zero number of overstatement misstatements.

Formula Method—Some Misstatements Expected

6.26 When planning a MUS sample, the auditor controls the risk that a material misstatement exists when, in fact, it does not (that is, risk of incorrect rejection) by making an allowance for expected misstatements in the sample. The auditor ordinarily specifies a desired allowance for sampling risk so that the estimate of projected misstatement plus the allowance for sampling risk will be less than or equal to tolerable misstatement.

6.27 If the auditor expects misstatements, and the auditor is not using the table approach (table 4-5), but using a formula approach with confidence factors described earlier, he or she may consult table C-2 in appendix C to identify an appropriate confidence factor that considers expected misstatement, and then

Monetary Unit Sampling

proceed to determine sample size using the same approach previously described when zero misstatements were expected.[8]

6.28 As an example of the method using confidence factors, an auditor using MUS might have assessed tolerable misstatement as $15,000 and the desired risk of incorrect acceptance as 5 percent. In addition, the auditor may expect approximately $3,000 of misstatement in the population to be sampled. The auditor would compute the ratio of expected to tolerable misstatement as 20 percent (that is, $3,000 ÷ $15,000). By reference to table C-2 in appendix C, the auditor locates the indicated confidence factor at the intersection of the risk of incorrect acceptance (5 percent) and the expected-to-tolerable ratio (20 percent). The confidence factor is 4.63.[9]

6.29 Using the formula approach, the confidence factor is divided by the tolerable misstatement percentage of the population of 0.03 (that is, $15,000 ÷ $500,000). The resultant sample size is 154.3 items, and is rounded up to 155 items. The sampling interval is computed to be $3,225 ($500,000 ÷ 155).[10]

6.30 Because MUS is based on attributes theory, yet another method is to refer directly to the statistical attribute sample size tables for tests of controls (see table A-1, "Statistical Sample Sizes for Tests of Controls—5 Percent Risk of Overreliance," in appendix A, "Attributes Statistical Sampling Tables"). This approach assumes a "worst case" scenario where any misstatements identified will be 100 percent misstatements, and thus may result in conservative sample sizes. Other MUS methodologies may allow for other assumptions about the average or maximum misstatement that might be found in order to refine the sample size. To use the tables in appendix A, the auditor converts the tolerable misstatement and the expected misstatement into percentages of the population's recorded amount and uses a sample size for the equivalent rates shown in the table. For example, if the auditor is designing a MUS sampling application for a population with a recorded amount of $500,000, he or she might have assessed tolerable misstatement as $15,000 and expected $2,500 of misstatement in the population. The auditor would calculate tolerable misstatement to be 3 percent ($15,000 ÷ $500,000) of the recorded amount and the expected misstatement to be 0.5 percent ($2,500 ÷ $500,000) of the recorded amount. The sample size for a 5 percent risk of concluding that controls are more effective than they actually are (see table A-1 in appendix A) is 157, where the tolerable misstatement is 3 percent and the expected misstatement rate is 0.5 percent. The auditor then determines the sampling interval to be $3,184 ($500,000 ÷ 157). If the auditor calculated a percentage of expected misstatement that is not shown on the table, he or she would usually interpolate in the table. In the example, if the expected misstatement was $3,000 (0.6 percent of the recorded amount), the appropriate sample size interpolated from table A-1 would be 178. The sampling interval would be $2,808 ($500,000 ÷ 178). Similarly, if the auditor were to calculate a percent for tolerable misstatement that is not shown on the table, he or she would interpolate the approximate sample size. The

---

[8] In the prior versions of this guide, another formula method using expansion factors was illustrated. That alternative method, with caveats regarding its use, is discussed further in table C-4, "Alternative MUS Sample Size Determination Using Expansion Factors," in appendix C, "Monetary Unit Sampling Tables."

[9] Interpolation can be used within the table for values that are not shown in the table.

[10] Note that the use of table 4-5, "Illustrative Sample Sizes," would result in the same sample size.

AAG-SAM 6.30

auditor then would calculate the sampling interval by dividing the recorded amount by the sample size.

6.31 In a particular situation the various sample size determination approaches can result in slightly different sample sizes.

Evaluating the Sample Results

6.32 The auditor using MUS projects the misstatement results of the sample to the population from which the sample was selected and calculates an allowance for sampling risk. If the entire sample is audited and no misstatements are found in the sample, the misstatement projection is zero dollars and the allowance for sampling risk is less than or equal to the tolerable misstatement used in designing the sample. If no misstatements are found in the sample, the auditor may conclude without making additional calculations that the recorded amount of the population is not overstated by more than the tolerable misstatement at the specified risk of incorrect acceptance.

6.33 If misstatements are found in the sample, the auditor calculates a projected misstatement and may calculate an allowance for sampling risk. This guide illustrates one means of calculating projected misstatement and an allowance for sampling risk that is appropriate for MUS samples selected using the method described in this chapter. The discussion of this method is limited to overstatements because the MUS approach is designed primarily for overstatements. If understatements are a significant consideration (in terms of expected number or percentage of book amount), the auditor ordinarily considers at the planning stage whether a separate MUS of a related population or an item-based classical sampling technique designed to detect understatements is appropriate.

6.34 MUS methodology for evaluating the effect of an overstated item takes into account whether it is 100 percent overstated or partially overstated when calculating the projected misstatement and an allowance for sampling risk.

Sample Evaluation With 100 Percent Misstatements

Projected Misstatement

6.35 A procedure to evaluate 100 percent misstatements identified in sample items is described in the following paragraphs. Because each selected dollar represents a group of dollars, the percentage of misstatement in the logical unit represents the percentage of misstatement or *tainting* for the whole sampling interval. For example, if the sampling interval is $5,000 and a selected account receivable with a recorded amount of $100 has an audit amount of zero dollars ($100 misstatement is 100 percent of the recorded amount), the projected misstatement is $5,000 (100 percent of $5,000). If the same account receivable had an audited amount of $30 ($70 misstatement is 70 percent of the recorded amount), the projected misstatement would be $3,500 (70 percent of $5,000). If a logical unit equals or exceeds the sampling interval, the projected misstatement is the actual amount of misstatement for the logical unit. The auditor adds the projected misstatements for all sampling intervals to calculate the total factual and projected misstatement for the population.

Upper Limit on Misstatement—100 Percent Misstatements Only

6.36 When evaluating a MUS sample statistically,[11] the auditor normally calculates an upper limit on misstatement equal to the projected misstatement found in the sample plus an allowance for sampling risk. The auditor may use either a computer program or a table of confidence factors as an aid in calculating the upper limit on misstatement. The first two columns shown in table 6-2, "Five Percent Risk of Incorrect Acceptance," are from table C-3, "Monetary Unit Sampling—Confidence Factors for Sample Evaluation," in appendix C.

Table 6-2

Five Percent Risk of Incorrect Acceptance

Number of Overstatements	Confidence Factor	Incremental Changes in Factor
0	3.00	—
1	4.75	1.75
2	6.30	1.55
3	7.76	1.46
4	9.16	1.40
5	10.52	1.36

6.37 The third column is the difference between the confidence factor for a specific number of overstatements and that of its predecessor.

6.38 If no misstatements are found in the sample, the upper limit on misstatements equals the confidence factor for no misstatements at a given risk of incorrect acceptance multiplied by the sampling interval.

Upper limit on misstatement = Confidence factor × Sampling interval

6.39 This upper limit when no misstatements are found, also referred to as *basic precision*, represents the minimum allowance for sampling risk inherent in the sample. For example, if the auditor specified a 5 percent risk of incorrect acceptance, used a $5,000 sampling interval, and found no misstatements, the upper limit on misstatements equals $15,000 (3 × $5,000). Because no misstatements are found, the projected misstatement is zero, and the allowance for sampling risk equals the upper limit on misstatement.

6.40 However, if two complete misstatements were found in the sample (for example, recorded accounts-receivable balances of $10 and $20 were each found to have an audited amount of zero), the auditor would calculate the upper limit on misstatement by multiplying the confidence factor for the actual number of misstatements found, at the given risk of incorrect acceptance, by the sampling interval. The upper limit is $31,500 (6.3 × $5,000). The $31,500 represents a projected misstatement of $10,000 (2 misstatements at 100 percent × $5,000) and, therefore, a precision (that is, allowance for sampling risk) of $21,500 ($31,500 − $10,000).

---

[11] Statistical calculations are not required for nonstatistical samples. Such calculations are normally required for a valid statistical application.

6.41 If the logical units in which the 100 percent misstatements occurred were equal to or larger than the sampling interval (for example, $15,000 and $20,000 instead of the $10 and $20 misstatements in the previous example), the upper limit on misstatement would equal (*a*) the factual misstatements in the logical units equal to or greater than the sampling interval, plus (*b*) the basic precision. Misstatements in items examined 100 percent or in items that equal or exceed the sampling interval do not increase the allowance for sampling risk. In this example, the upper limit would equal $35,000 ($15,000 + $20,000) plus $15,000 (3 × $5,000), or a total of $50,000. The auditor is required to accumulate this result with the misstatements discovered in any other items examined 100 percent.

Sample Evaluation With Less Than 100 Percent Misstatements

6.42 In many sampling applications, the auditor identifies misstatements in which the logical unit is not completely incorrect. In these situations, the *tainting* (misstatement percent) is less than 100 percent.

Projected Misstatement When Taintings Occur

6.43 To project misstatements when taintings occur, one approach is to determine the percentage of misstatement in the logical unit and multiply this percentage by the sampling interval. For example, if a receivable balance with a recorded amount of $100 has an audit amount of $50, the auditor would calculate a 50 percent tainting ($50 ÷ $100). A tainting percentage is calculated for all logical units with misstatements except those that have recorded amounts equal to or greater than the sampling interval. The auditor may then multiply the tainting percentage by the sampling interval to calculate a projected misstatement. By adding the sum of all projected misstatements to the actual misstatement found in the logical units equal to or greater than the sampling interval, the auditor calculates the total factual and projected misstatement. For example, 6 misstatements might have been identified in the sample. Table 6-3, "Calculation of Total Projected Misstatement," shows how the auditor calculated the total projected misstatement.

Table 6-3

Calculation of Total Projected Misstatement

A	B	C	D	E	F
Recorded Amount	Audit Amount	Factual Misstatement (A − B)	Tainting (C ÷ A)	Sampling Interval	Projected Misstatement (D × E)
$100	$25	$75	75%	$5,000	$3,750
1,000	950	50	5%	5,000	250
500	250	250	50%	5,000	2,500
50	0	50	100%	5,000	5,000
10	9	1	10%	5,000	500
10,000	9,000	1,000	N/A[1]	N/A[2]	1,000
Total Factual and Projected Misstatement					$13,000

[1] The logical unit is greater than the sampling interval; therefore, the projected misstatement equals the actual misstatement. Some auditors remove all items in excess of tolerable misstatement from the population before sampling, to reduce the complexity of the sample evaluation.

[2] See table note 1.

Upper Limit on Misstatements When Taintings Occur

6.44 The allowance for sampling risk when taintings occur includes both the *basic precision* and an incremental allowance resulting from the occurrence of misstatements. To calculate that incremental allowance, the auditor divides the misstatements into two groups: (*a*) those occurring in logical units less than the sampling interval and (*b*) those occurring in logical units equal to or greater than the sampling interval. In the preceding example, the first five misstatements are in the first group, and the last misstatement is in the second group.

6.45 Misstatements occurring in logical units equal to or greater than the sampling interval have no allowance for sampling risk associated with them because all logical units of this size have been examined. Sampling risk exists only when sampling takes place.

6.46 One conservative approach[12] to calculating the allowance for sampling risk is to rank the projected misstatements by percentage of tainting in descending order and then calculate the incremental allowance for sampling risk for each misstatement. This is done by (*a*) multiplying the projected misstatement for each misstatement occurring in a logical unit that is less than the

---

[12] The upper limit that results from the approach illustrated here is known as the *Stringer Bound*, after Kenneth J. Stringer. Other methods such as the cell method have been shown to be effective in simulation studies.

sampling interval by the incremental change in the confidence factor and (*b*) subtracting the related projected misstatement. In the preceding example, the auditor could rank the estimates of misstatements as shown in table 6-4, "Calculating the Allowance for Sampling Risk." The $19,253 represents $12,000 in projected misstatement and $7,253 in additional allowance for sampling risk.

Table 6-4

Calculating the Allowance for Sampling Risk

Projected Misstatement	Incremental Changes in Confidence Factor	Projected Misstatement Plus Incremental Allowance for Sampling Risk
$5,000	1.75	$8,750
3,750	1.55	5,813
2,500	1.46	3,650
500	1.40	700
250	1.36	340
$12,000		$19,253

6.47 To calculate the upper limit on misstatement, the auditor adds the $19,253 to 2 components: (*a*) the *basic precision* and (*b*) the misstatements, if any, occurring in logical units equal to or greater than the sampling interval. In the example, the *basic precision* was calculated to be $15,000 (3 × $5,000) and the misstatement occurring in logical units equal to or greater than the sampling interval is $1,000. The upper limit on misstatement is $35,253 ($19,253 + $15,000 + $1,000).

6.48 The sample results can be summarized as follows:

a. The sample contains factual misstatement of $1,426.
b. The total factual and projected misstatement is $13,000.
c. The total precision (allowance for sampling risk) is $22,253 (basic precision of $15,000 plus $7,253 incremental allowance for sampling risk).[13]
d. Therefore, there is a 5 percent risk that the recorded amount is overstated by more than $35,253.

Quantitative Considerations

6.49 Usually, if the upper limit on misstatements is less than the tolerable misstatement, the sample results will support the conclusion that the population is not misstated by more than tolerable misstatement at the specified risk of incorrect acceptance. If the upper limit on misstatement exceeds tolerable misstatement, the sample results do not support the conclusion that the population is not misstated by more than tolerable misstatement. This result might have been obtained because the rate and size of misstatements exceeded the auditor's expectation of misstatement, or because the sample size

---

[13] Alternatively, it is the upper limit on the misstatement of $35,253 minus the projected misstatement of $13,000.

was too small to support the desired assurance, given the misstatements found. In designing a MUS application, the auditor makes an assumption about the amount of misstatement in the population. If the sample results do not support the auditor's expectation of misstatement because more misstatement exists in the population than was expected, the allowance for sampling risk will not be adequately limited. The auditor may

    a. examine an additional representative sample from the chosen population if the auditor determines that extending the sample is appropriate. Because of the mechanics of MUS, many auditors use an additional number of sampling units equal to or greater than the original sample size.[14] Before extending the sample, auditors are reminded that selecting and auditing additional sample items will often reveal similar or more misstatement than the original sample.

    b. perform additional substantive procedures directed toward the same audit assertion. This reliance on other tests would allow the auditor to accept a greater risk of incorrect acceptance for the sampling application. Recalculating the allowance for sampling risk with the greater risk of incorrect acceptance will not change the projected misstatement (point estimate) of the population, but it will decrease the upper limit on the misstatement. Typically, this approach may be effective only when differences between the desired and achieved results are small, because other tests such as analytical procedures may not provide the quality of evidence regarding the population that a sample might provide.

6.50 Occasionally, the sample results might not support acceptance of the recorded amount, because although the auditor selects a sample that is expected to be unbiased (representative of the population), the sample selected might not be representative of the population. This happens because sample results can differ due to the different potential combination of sample items that can be selected from a population. If all the related audit evidence contradicts the sample evidence, the auditor might suspect that the sample is not representative of the population. Additionally, if an analysis of the sample items, and the specific misstatements identified supports that suspicion, the auditor may choose to examine additional sampling units or perform alternative procedures to determine whether the recorded amount of the population is misstated. A greater level of misstatement observed in the sample result than expected is usually not sufficient support that the sample is not representative.

6.51 If the sample results do not support the recorded amount of the population and the auditor believes the recorded amount is misstated, the auditor typically considers that information as well as other audit evidence when evaluating whether the financial statements as a whole may be materially misstated. In this situation, the auditor would request that the entity correct the factual misstatements and investigate the underlying circumstances contributing to the projected misstatements and, if appropriate, adjust the recorded amount.

---

[14] To select a sample in this circumstance, the auditor may divide the original sampling interval in half and, using the resulting sum, begin selecting the expanded sample by using the same random start. If that random start exceeds the new sampling interval, the auditor subtracts the new sampling interval from the original random start. This results in a sample consisting of the original sample plus additional sampling units. The complexities of alternative methods of expanding the sample are beyond the scope of this guide and may require the assistance of a statistical sampling specialist.

After adjustment, if the upper limit on misstatement is less than the tolerable misstatement, the sample results would support the conclusion that the adjusted population is not misstated by more than tolerable misstatement at the specified risk of incorrect acceptance.

Qualitative Considerations

6.52 In addition to evaluating the frequency and amounts of monetary misstatements, the auditor should consider the qualitative aspects of misstatements, pursuant to paragraph .11 of AU-C section 450, *Evaluation of Misstatements Identified During the Audit* (AICPA, *Professional Standards*). These considerations are discussed in chapter 4.

MUS Sampling Case Study

6.53 Thaddeus Andrews of Andrews, Baxter & Co., is the auditor of the EZ Credit Bank, a privately owned commercial bank. Andrews established overall materiality at planning at $100,000 and a performance materiality of $75,000. Andrews designed a sampling application to test the existence, gross valuation, and accuracy assertions for EZ Credit's commercial loans-receivable balance as of September 30, 20XX. The balance of commercial loans receivable was $5 million as of September 30, 20XX. Andrews expected little, if any, misstatement to exist in the relevant assertions in the commercial loans-receivable balance because of the bank's strong control environment and effective controls over loan transactions. If any misstatements did exist, Andrews believed that they would be overstatements. As a result, Andrews decided that MUS would be an appropriate sampling approach to use. Because of the strong controls and because of the importance of controls in the banking industry, Andrews decided to test controls as a basis for assessing control risk (and *risks of material misstatement*) as low. He also decided to place moderate reliance on analytical procedures and other substantive tests.

6.54 Andrews decided to confirm all selected commercial loans receivable with the bank's customers. In making decisions about performance materiality and tolerable misstatement, Andrews considered the factors in table 4-3, "Factors to Consider in Setting Performance Materiality (PM) at the Engagement Level and Tolerable Misstatement (TM) at the Test Level." He believed that a misstatement of $55,000 or more in the commercial loans-receivable balance, when combined with misstatements from other tests of this balance and misstatements in other accounts that might be found, could result in amounts exceeding materiality or an inadequate allowance for sampling risk, or both. As a result, he set the tolerable misstatement for this sampling application at $55,000. Because all the relevant factors in table 4-3 were considered in the determination of performance materiality, Tolerable misstatement was set at the performance materiality amount in this circumstance.[15] He documented the factors considered supporting the assessment of the performance materiality (and tolerable misstatement) amounts. Because Andrews assessed control risk and the risk of material misstatement as low and performed a number of moderately effective analytical procedures to test the commercial loans

---

[15] If only some of the relevant factors had been considered when determining performance materiality, then performance materiality might have been set higher than $55,000 (set closer to materiality), and the tolerable misstatement for this test would then be lower than the performance materiality.

receivable, he determined that a 37 percent risk of incorrect acceptance (63 percent confidence) was appropriate for the confirmation sample.

6.55 Andrews assumed some misstatement in the account balance when calculating the appropriate sample size. He used an expected misstatement of $11,000 when he designed his sampling application. Although this resulted in a somewhat larger sample size, planning to find some misstatement when determining the sample size also reduced the possibility that he would have to extend the sampling application or perform other procedures if the misstatements found exceeded his expectations.

Selecting the Sample

6.56 Andrews calculated the appropriate sample size and sampling interval as follows:

Tolerable Misstatement	$55,000
Expected Misstatement	$11,000
Ratio of Expected to Tolerable Misstatement	0.20
Tolerable Misstatement ÷ Population	0.011
Confidence Factor From Table C-2	1.3
Confidence Factor ÷ Tolerable Percentage = Sample Size (Rounded Up)	119

6.57 Andrews then calculated the sampling interval of $42,016[16] by dividing the recorded amount of the commercial loans receivable by the sample size ($5,000,000 ÷ 119). Andrews did not need to identify the commercial loans that individually exceeded materiality, performance materiality, or tolerable misstatement, as there were none, and decided to allow the population to include any items greater than the tolerable misstatement of $55,000 because the systematic selection method he used would be certain to select all logical units with recorded amounts greater than or equal to the $42,016 sampling interval. Andrews used computer software to systematically select his sample.

6.58 The selected sample included 116 customer balances rather than the 119 originally calculated because 3 accounts were larger than $42,016 and were automatically included in the items to be examined 100 percent.

Evaluating the Sample Results

6.59 Andrews mailed confirmation requests to each of the 119 (116 + 3) customers whose commercial loan balances had been selected. Of the confirmation requests, 90 were completed and returned to him. Andrews was able to obtain reasonable assurance through alternative procedures that the remaining 29 balances were bona fide receivables and were not misstated. Of the 90 responses, only 2 indicated that the recorded balances were overstated.

6.60 Andrews calculated the projected misstatement as shown in table 6-5, "Andrews's Calculation of Projected Misstatement."

---

[16] In practice, the interval may be rounded down to a more convenient numerical value such as $42,000.

Table 6-5

Andrews's Calculation of Projected Misstatement

	A	B	C	D	E	F
Misstatement Number	Recorded Amount	Audit Amount	Misstatement (A − B)	Tainted (C ÷ A)	Projected Sampling Interval	Projected Misstatement (D × E)
1	$9,000	$8,100	$900	10%	$42,016	$4,202
2	500	480	20	4%	$42,016	1,681
Projected misstatement						**$5,883**

6.61 He then calculated an allowance for sampling risk (precision). The allowance consisted of two parts: the basic precision and the incremental allowance.

Sampling interval	$42,016
Multiplied by confidence factor for a 37 percent risk of incorrect acceptance	× 1.00
Basic precision	$42,016

6.62 The incremental allowance was calculated as follows:

Misstatement Number	Projected Misstatement	Incremental Factor[17]	Projected Misstatement × Incremental Factor
1	$4,202	1.14	$4,790
2	1,681	1.11	1,866
	$5,883		$6,656
Less projected misstatement			5,883
Incremental allowance			$733

6.63 Andrews compared the upper limit on the misstatement (that is, projected misstatement[18] plus an allowance for sampling risk) of $48,632 ($5,883 + $42,016 + $733) with the tolerable misstatement of $55,000. Because the upper limit on misstatement was less than tolerable misstatement, he concluded that the sample results supported the conclusion that the recorded amount of the commercial loans receivable was not materially misstated with respect to the assertions relevant to this test. Andrews also concluded that the overstatements were due to ordinary misstatements in the accounting process and that they did not require him to modify his planned substantive procedures or his assessment of the risks of material misstatement; however, the "best

---

[17] These factors are for the 63 percent level of confidence.
[18] Had Andrews identified factual misstatement in the items examined 100 percent, that misstatement would have been added to the projected misstatement (including an allowance for sampling risk) when calculating the upper limit on the misstatement.

estimate"[19] of the sample indicated a projected $5,883 overstatement, and he aggregated the projected misstatement from the sample results with other factual and projected (likely) misstatements[20] when he evaluated whether the financial statements as a whole were materially misstated. He brought the factual and projected misstatements to the attention of management and those charged with governance. They decided not to make any adjustment except for the amounts of factual misstatement.

---

[19] Also termed the *point estimate* or *direct projection* of the misstatement.

[20] In addition, any *judgmental differences* would also be accumulated. Judgmental differences arise from judgments of management concerning accounting estimates that the auditor considers unreasonable or the selection or application of accounting policies that the auditor considers inappropriate.

Chapter 7

Classical Variables Sampling

> **⊙ Update 7-1 *Audit*: Clarified Auditing Standards**
>
> The auditing guidance in this guide edition has been conformed to Statement on Auditing Standards (SAS) Nos. 122–125, which were issued in 2011 as part of the Auditing Standards Board's Clarity Project. These clarified SASs are effective for audits of financial statements for periods ending on or after December 15, 2012. Although extensive, the revisions to generally accepted auditing standards resulting from these clarified SASs do not change many of the requirements found in the auditing standards that they supersede.
>
> To assist auditors and financial reporting professionals in making the transition, this guide includes appendix F, "Mapping and Summarization of Changes—Clarified Auditing Standards," which provides a cross reference of the sections in the superseded auditing standards to the applicable sections in the clarified auditing standards and identifies the changes, either substantive or primarily clarifying in nature, that may affect an auditor's practice or methodology relative to the applicable sections of SAS Nos. 122–125. It also summarizes the changes resulting from the requirements of SAS Nos. 122–125.
>
> The preface of this guide and the Financial Reporting Center on www.aicpa.org provide more information on the Clarity Project. Visit www.aicpa.org/sasclarity.

7.01 This chapter describes several classical variables sampling techniques and some of the factors to be considered by an auditor applying these techniques.

7.02 Classical variables sampling techniques use normal distribution theory to evaluate selected characteristics of a population on the basis of a sample of the items constituting the population. The design of a classical variables sampling approach involves mathematical calculations that tend to be complex and difficult to apply manually. Because auditors generally use computer programs to assist them in determining sample sizes and evaluating sample results for classical variables sampling applications, it is not essential for auditors to know mathematical formulas to use these methods. Consequently, such formulas are not provided in this guide. These formulas are readily available in numerous books that deal with sampling theory.

Selecting a Statistical Approach

7.03 Both statistical approaches to sampling for substantive testing—classical variables sampling and monetary unit sampling (MUS)—can provide sufficient evidential material to achieve the auditor's objective; however, in a given circumstance one might be more appropriate than the other.

AAG-SAM 7.03

Advantages

7.04 The advantages of classical variables sampling include the following:

- If there are many differences between recorded and audited amounts, classical variables sampling might meet the auditor's objectives with a smaller sample size.

- Because most classical variables samples are selected on an item, and not a proportional to size basis, they are often the most appropriate techniques for sampling populations where understatements are the focus or a concern.

- Classical variables samples may be easier to expand if that becomes necessary by selecting additional sample items for each of the strata without reordering the population and creating a second probability proportional to size (PPS) selection.

- Inclusion of zero value items in the population for possible selection in the sample generally does not require special sample design considerations.[1] If examining zero value items is important to the auditor's objectives, the auditor using MUS designs a separate test of zero amount items, because the PPS method of sample selection described in this guide would not select zero valued items.

- Inclusion of negative value items in the evaluation of a classical variables sample generally does not require special sample design considerations.[2] A MUS sample might need to be designed with special considerations to include negative items in the sample evaluation.

Disadvantages

7.05 The disadvantages of a classical variables sampling approach include the following:

- Classical variables sampling is more complex than MUS. Generally, an auditor needs the assistance of computer programs to design a classical variables sample, select the sample, and evaluate sample results.

- To determine a sample size for a classical variables sample, the auditor generally needs an estimate of the standard deviation of the characteristic of interest in the population. Because the auditor generally does not know this information when designing a sample, he or she determines the appropriate sample size based on an estimate of this standard deviation. This estimate might be difficult to make. In some applications, if the population is maintained on a computer file and the auditor is able to analyze the file using computer-assisted audit techniques, he or she may be able to measure the standard deviation of the recorded amounts as a reasonable estimate of the standard deviation of the audited

---

[1] However, such items may have different audit and risk implications that require special consideration, and thus may require these items to be segregated and examined separately.

[2] See footnote 1.

- amounts or characteristic of interest (such as the difference between the recorded and audited amount). This estimate may also be based on the standard deviation of a pilot sample or the auditor's prior knowledge of the population.
- When there are (*a*) either very large items or very large differences between recorded and audited amounts in the population and (*b*) the sample size is small, the normal distribution theory[3] may not be appropriate. As a result, the auditor might accept an unacceptable recorded amount of the population more often than the desired risk of incorrect acceptance. In addition when misstatements are rare, some classical variables sampling techniques such as the difference and ratio techniques are not able to be applied.
- Classical variables sampling techniques may be applied to an account because it might contain understatements. When misstatements are not expected or are expected to be rare, classical variables sampling techniques that are based on finding an adequate representation of differences (for example, difference or ratio methods) may not be practical. In such cases, some auditors apply MUS and perform other tests (such as analytical procedures, selections from related populations, or control tests) to determine whether there is a risk that understatements were not detected.

7.06 The auditor typically considers the advantages and disadvantages of classical variables sampling versus MUS when deciding which approach to use. Some applications in which a classical variables approach may be especially useful include the following:

- Inventory test counts and price tests in which the auditor anticipates a significant number of audit differences between audited and recorded amounts or where both overstatements and understatements are likely to exist
- Testing the underlying data associated with estimates of allowance or valuation amounts where overstatements or understatements are equally likely
- Conversion of inventory from first in, first out to last in, first out
- Applications for which the objective is to estimate independently the total amount of the population

Types of Classical Variables Sampling Techniques

7.07 There are three classical variables sampling methods discussed in this chapter: the mean-per-unit, difference, and ratio approaches. Another technique that is related to the ratio technique, but not described in this chapter, is the regression estimator. Despite its more complex computations, it may perform better in some circumstances than the ratio estimation or difference estimation methods.

---

[3] Various correction factors such as use of the Student T distribution or use of a finite population correction factor may extend the usefulness of classical techniques in smaller samples and populations. Auditors sometimes use minimum sample sizes to overcome issues related to small sample sizes.

AAG-SAM 7.07

Mean-Per-Unit Approach

7.08 When using this approach, the auditor estimates a total population amount by calculating an average audited amount for all items in the sample and multiplying that average amount by the number of items constituting the population. For example, an auditor has randomly selected 200 items from a population of 1,000 inventory items. After determining the correct purchase price and recalculating price-quantity extensions, the auditor determines the average audited amount for items in the sample by totaling the audited amounts of the 200 sampling units and dividing by 200, which equals $980. The estimated inventory balance is then calculated as $980,000 ($980 × 1,000). Using normal distribution theory based on the variability (that is, standard deviation) of the audited amounts in the sample, the auditor also calculates an allowance for sampling risk for a specified risk of incorrect acceptance.

Difference Approach

7.09 When using this approach, the auditor calculates the average difference between audited and recorded amounts of the sample items and projects that average difference to the population. For example, an auditor has examined 200 items from a population of 1,000 inventory items. The total recorded amount for the population is $1,040,000. The auditor compares the audited amount with the recorded amount for each of the 200 sampling units and accumulates the difference between the recorded amounts ($208,000) and the audited amounts ($196,000)—in this case, $12,000. The difference of $12,000 is divided by the number of sample items (200) to yield an average difference of $60. The auditor then multiplies the average difference by the number of items in the population to calculate a total difference of $60,000 ($60 × 1,000) between the recorded amount and audited amount. Because the total recorded amount of the sampling units is greater than the total audited amount, the difference is subtracted from the total recorded amount to obtain an estimated inventory balance of $980,000.[4] The auditor also calculates an allowance for sampling risk using normal distribution theory based on the variability (that is, standard deviation) of the differences between the recorded amount and the audited amount of the sampling units for a specified risk of incorrect acceptance.

Ratio Approach

7.10 When using this approach, the auditor calculates the ratio between the sum of the audited amounts and the sum of the recorded amounts of the sample items and projects this ratio to the population. The auditor estimates the total population amount by multiplying the total recorded amount for the population by the same ratio. If the auditor had used the ratio approach in the previous example, the ratio of the sum of the sample's audited amounts to the sum of the sample's recorded amounts would have been 0.94 ($196,000 ÷ $208,000). The auditor would multiply the total recorded amount for the population by this ratio (0.94) to obtain an estimate of the inventory balance of $978,000 ($1,040,000 × 0.94). The auditor would also calculate an allowance for sampling risk using normal distribution theory based on the extent and magnitude of the differences for a specified risk of incorrect acceptance.

---

[4] It should be noted that in practice, the use of the mean and difference approaches would not often result in the exact same projected amount.

Choosing a Classical Variables Sampling Approach

7.11 Chapter 4, "Nonstatistical and Statistical Audit Sampling for Substantive Tests of Details," provided the general considerations in using audit sampling for substantive tests. This section describes additional factors the auditor considers when using classical variables sampling for a substantive test.

The Ability to Design a Stratified Sample

7.12 As discussed in chapter 4, the auditor can often reduce sample size by effectively stratifying a population. Stratification is usually necessary whenever classical variables sampling is applied. For example, an unstratified mean-per-unit approach requires sample sizes that may be too large to be efficient for ordinary audit applications. Nevertheless, there are circumstances, however, when the auditor might efficiently use an unstratified mean-per-unit sampling approach. For example, stratification might not be necessary in a population of items of similar size and risk. Mean-per-unit may be the only technique available when the recorded amounts of the individual items are not available, cannot be matched with units such as after a loss of records, or are not at all reliable. When samples include enough misstatements, difference and ratio estimators are often more efficient and effective estimators than the mean-per-unit approach.

The Expected Number of Differences Between the Audited and Recorded Amounts

7.13 Both the ratio and the difference approaches require that sufficient differences between the audited and recorded amounts exist in the sample. If no differences exist between the audited and recorded amounts of the sample items, the mechanics of the formula underlying each of these methods leads to the erroneous conclusion that the allowance for sampling risk[5] is zero—that is, there is no sampling risk. Such a conclusion is erroneous because sampling risk always exists unless the auditor examines all items constituting the population. There is no hard and fast rule about how many differences are necessary to estimate accurately the allowance for sampling risk for a sample using either the ratio or difference approach. A minimum of 20 or more differences is generally suggested. When stratified sampling is used, these techniques may also require a minimum number of differences be found per stratum in order to make the statistical computations. Failure to find the required number of differences per stratum may require the combination of strata in the evaluation of the sample results, If the auditor decides to use a statistical approach and expects to find only a few or no differences, he or she considers whether alternative approaches such as mean-per-unit or MUS would be more appropriate, or considers engaging a sampling specialist to assist in the analysis.

Required Information

7.14 In addition to sample size, all the classical variables approaches require different information for the population or for each stratum, if stratified

---

[5] Allowance for sampling risk (that is, precision) is a measure of uncertainty around the sample projection. All samples by nature are subject to some sampling risk. To have zero sampling risk, all items in the population would have to be examined.

sampling is used. To use the mean-per-unit approach, the auditor needs to know the total number of items in each stratum and an audited amount for each sampling unit. Both the ratio and the difference approaches require an audited amount and recorded amount for each sampling unit. The recorded amount may be developed from the entity's normal recordkeeping system (for example, the inventory shown by the perpetual records), or it may be any amount developed by the entity for each item in the population (for example, the entity's priced inventory). In both approaches the auditor needs to know the recorded amount for the total population and the total number of items in the population. Additionally, the auditor will generally consider whether the entity has properly accumulated the recorded amounts of the items in the population (for example, checked for duplicate sampling units, omissions of sampling units, and so on) when the sample item recorded amount is used in the computation.

7.15 Depending on the circumstances, many auditors prefer to use either the difference or the ratio approach. These methods are generally more efficient than the mean-per-unit approach because the difference and the ratio procedures provide projections directly of the misstatements found in the sample and generally require smaller sample sizes to achieve the same confidence (risk of incorrect acceptance) and precision (allowance for sampling risk). The more information an auditor has about the population and the sampling units, the greater his or her ability to design an efficient sample.

Determining the Sample Size

7.16 Sample size depends on the variability of the characteristic of audit interest, by stratum for stratified samples, tolerable misstatement, and the acceptable risk of incorrect acceptance. Because auditors usually use computer programs to determine appropriate sample sizes for classical variables sampling applications, they generally do not need to apply the mathematical formulas to use these methods; however, knowledge of the assumptions and computational routines can assist auditors in understanding these methods and using projection methods that are most appropriate for the sample results obtained.

Considering Variation Within the Population

7.17 Chapter 4 discussed the effect variation in the population had on sample size. The sample size required for a classical variables sampling application increases as the variation (measured by the standard deviation) becomes greater. In general, any change in the variation in the population affects the sample size by the square of the relative change. For example, the unstratified sample size for a given risk of incorrect acceptance, population size, tolerable misstatement, and amount of variation in the population has been determined to be 100. If the amount of variation was twice the original amount, the sample size necessary to meet the auditor's objectives would be 4 multiplied by the original sample size (in this case, a sample size of 400). To the extent that stratification reduces standard deviation, it can have a significant impact on sample size and efficiency.

7.18 The optimal number of strata depends on the circumstances. After a certain point, division of the population into additional strata has a diminishing effect on the variation within each stratum and adds complexity and cost. The auditor may consider the additional costs of dividing the population into more strata in relation to the resulting reduction of the overall sample size. A general rule of thumb often followed is that between 3 and 10 strata are often effective and efficient. The need to have some minimum number of sample items or differences in each stratum (not tested 100 percent) for proper analysis often makes a larger number of strata impractical.

7.19 Stratification can be performed on computerized records with the assistance of programs designed for such audit applications. Stratification is more time-consuming and may be impractical when the auditor has to select the sample manually. In some circumstances, auditors subjectively determine strata boundaries based on their knowledge of the population's composition. Some auditors believe it is usually not efficient to manually divide a population, after removing the items to be examined 100 percent, into more than 2 or 3 strata. In those cases, the auditor then estimates the variation for each stratum, uses the tolerable misstatement and risk of incorrect acceptance for the population, to calculate the sample size, and allocates a portion of the sample size to each stratum. Certain populations (for example, student loans, certain awards and grants, or loans for a specific purpose) may be sufficiently similar in size or in expected misstatement differences or ratios so that stratification is not essential.

Calculating the Sample Size

7.20 Auditors consider tolerable misstatement, a measure of variance, and the risk of concluding that a material misstatement does not exist, when it does (risk of incorrect acceptance) when determining sample size.[6] In addition, they may also find it practical to consider explicitly the risk of incorrect rejection. Some computer programs for classical variables sampling applications allow the auditor to specify these factors when calculating a sample size. In controlling for this risk, the auditor needs to specify a confidence level associated with the risk of incorrect rejection as well as a confidence level for the risk of incorrect acceptance. Other computer programs do not have the functionality to allow the auditor to directly specify the two risks (incorrect acceptance and incorrect rejection). When this is the case, the auditor can determine an adjusted allowance for sampling risk (precision) by relating the tolerable misstatement and the risk of incorrect acceptance to a given level of the risk of incorrect rejection. Table D-1, "Ratio of Desired Allowance for Sampling Risk of Incorrect Rejection to Tolerable Misstatement," in appendix D, "Ratio of Desired Allowance for Sampling Risk of Incorrect Rejection to Tolerable Misstatement," illustrates the relationship of these factors that can be used to determine an appropriate desired allowance for sampling risk that will provide the specified protection against incorrect acceptance. Not all software programs use the same terminology as this guide, and users are advised to understand how the requested program inputs relate to the concepts in this guide.

7.21 In planning a one-sided classical variables sampling application, for example, the auditor might wish to specify a tolerable misstatement of

---

[6] Expected misstatement, a common sampling parameter (see paragraph .A13 of AU-C section 530, *Audit Sampling* [AICPA, *Professional Standards*]), is not used directly in the sample size calculation for a classical variables sample, but an estimate of the frequency and size of expected misstatements may nevertheless assist the auditor in assessing the potential variability, setting a precision for the sample, and selecting an appropriate classical variables sampling technique (for example, mean per unit, difference, or ratio technique).

$10,000, a 5 percent risk of incorrect acceptance, and a 10 percent one-sided risk of incorrect rejection. The auditor can plan a sample to achieve these dual objectives by setting the desired allowance for (sampling) risk of incorrect rejection (also known as the *precision*) at an appropriate fraction of tolerable misstatement read from table D-1 in appendix D.[7] This table shows that to achieve a 5 percent risk of incorrect acceptance and a 10 percent one-sided risk of incorrect rejection, the ratio of desired allowance for risk of incorrect rejection to tolerable misstatement should be 0.437. Accordingly, the auditor would set the desired allowance for risk of incorrect rejection at $4,370 ($10,000 × 0.437).

7.22 Although it depends on the specific software, it is common for classical variables sampling computer programs that calculate sample sizes to require the auditor to enter the risk of incorrect rejection (for example, 10 percent),[8] and the desired allowance for risk of incorrect rejection (for example, $4,370). If the auditor determines the desired allowance for risk of incorrect rejection from table D-1, the sample size should be sufficient to also achieve the desired risk of incorrect acceptance (for example, 5 percent) relative to tolerable misstatement (for example, $10,000).

7.23 The size of the sample required to achieve the auditor's objective is affected by changes in his or her allowance for sampling risk. The sample size required to achieve this at a given risk of incorrect rejection for a given population increases as the auditor specifies a smaller desired allowance for sampling risk. In general, any change in the desired allowance for sampling risk affects the sample size by the square of the relative change. For example, the sample size for a given desired allowance for sampling risk may be 100. If this allowance for sampling risk is reduced by one-half, the sample size would be 4 multiplied by the original sample size.

7.24 To protect against the possibility that the classical variables sampling methods might not yield appropriate sample sizes in some cases, some auditors use rules of thumb concerning minimum sample sizes for classical variables samples. For example, a homogeneous population (that is, the population comprises loans of a similar face amount) may result in an inappropriately small sample size computation due to the lack of variability in the recorded amounts. One rule of thumb is to set the minimum sample size (by stratum and in total) equal to what would have been selected using the MUS approach described in chapter 6, "Monetary Unit Sampling," assuming no misstatements are expected. Another rule of thumb is to establish minimum sample sizes for the overall application and per stratum, for example, 50–75 sampling units per application and a minimum of 20–30 sample items per stratum. The auditor or the audit software would then add additional items to the computed sample sizes for the strata to meet the minimums.

---

[7] If the auditor desires a sample that provides two-sided risk protection for risks of incorrect acceptance or incorrect rejection, the auditor would make an appropriate adjustment when using table D-1, "Ratio of Desired Allowance for Sampling Risk of Incorrect Rejection to Tolerable Misstatement." For example, to obtain a ratio for a 10 percent two-sided risk of incorrect rejection, the auditor would use the 5 percent risk of incorrect rejection column (in other words, the one-sided risk divided by 2).

[8] Many programs require the complement of this risk (in this example, 90 percent) to be entered, and may describe it as the *confidence level* (often a two-sided interval).

Evaluating the Sample Results

7.25 Each of the classical variables approaches to sampling provides the auditor with an estimated amount of the account balance or class of transactions being examined. As indicated previously, the difference between this estimated amount and the entity's recorded amount is the projected misstatement. Each approach also provides the auditor with an allowance for sampling risk (also referred to as *achieved precision*).

7.26 When it is unclear which evaluation approach is most consistent with the observed sample results and available computer programs, auditors may choose the technique that provides the smallest allowance for sampling risk, as that technique will often be the best one to evaluate the sample data.

7.27 Paragraph .14 of AU-C section 530, *Audit Sampling* (AICPA, *Professional Standards*), states "the auditor should evaluate the results of the sample, including sampling risk, and whether the use of audit sampling has provided a reasonable basis for conclusions about a population that has been tested."[9] This can often be achieved in MUS samples or nonstatistical samples by comparing projected misstatement to tolerable misstatement or comparing projected misstatement to the expected misstatement used in determining the sample size. In the case of classical sampling techniques where expected misstatement may not be used in determining the sample size, other sampling parameters such as measures of variability may provide evidence that the characteristics observed in the sample were properly considered when the sample was designed. If the entity records adjustments to the population, the point estimate and the upper limit are both reduced by the amount of the adjustment. The comparison of the remaining projected misstatement with tolerable misstatement and the consideration of (post adjustment) sampling risk are generally considered together when the auditor evaluates the results of a classical variables sample.

7.28 Because providing for a desired allowance for sampling risk related to the risk of incorrect rejection is a planning concept, the sample evaluation decision process uses the risk of incorrect acceptance and the tolerable misstatement (rather than the desired allowance for sampling risk of incorrect rejection determined from using table D-1).

7.29 For example, an auditor has calculated a sample size based on a 5 percent risk of incorrect acceptance and a 10 percent one-sided risk of incorrect rejection. The auditor has assessed tolerable misstatement to be $10,000 for a population with a recorded amount of $150,000 and has used a desired allowance for sampling risk of incorrect rejection of $4,370 for planning purposes to determine a sample size that should achieve the desired risks of incorrect acceptance and incorrect rejection (see appendix D). The auditor would use a 5 percent risk of incorrect acceptance and tolerable misstatement of $10,000 in evaluating the results.

7.30 When evaluating the sample results, assume the direct projection of the sample misstatement after applying audit procedures to the sample items is $5,000. The estimated population is $145,000. Thus the estimation of the lower limit of the population is $142,000, or $8,000 ($5,000 projected misstatement plus $3,000 allowance for sampling risk) less than the recorded

---

[9] In this context, the "reasonable basis" may be viewed as whether the sample has achieved the desired precision at the level of sampling risk used in planning the sample.

amount. Because this difference is less than tolerable misstatement ($10,000), the auditor may conclude that the sample supports that the population is not materially misstated.

7.31 If the difference between the recorded amount ($150,000, in the example) and the far end of the range from the sample ($142,000, in the example) were greater than tolerable misstatement ($10,000, in the example), the sample would not support the absence of a material misstatement at the risk level used in the evaluation.[10] In that case, the sample results might have been obtained due to one of the following reasons:

- The recorded amount was misstated by an amount greater than tolerable misstatement.
- The sample results yielded an allowance for sampling risk larger than desired by the auditor (for example, by underestimating the variability in the population) resulting in a sample size that was too small to give sufficiently precise results.
- The sample was not representative of the population.

7.32 However, suppose, in this example, the audit estimate of the population (based on a classical variables sample) is $145,000, with an allowance for sampling risk of $15,000 (that is, $145,000 minus $15,000 in possible overstatement). Because the difference between the recorded amount ($150,000) and the far end of the range ($130,000) is greater than the tolerable misstatement of $10,000, the sample results would not usually support acceptance of the recorded amount at the level of risk used in the design and evaluation of the sample.

7.33 If the variation of the characteristic of interest exceeds the auditor's estimate, the sample results might not adequately limit the allowance for sampling risk. Generally, the auditor using a computer program to perform a classical variables application can ascertain if this has occurred by comparing the standard deviation used to determine sample size with the standard deviation calculated as part of the evaluation of the sample results. When evaluating the sample results, if the standard deviation calculated is greater than the standard deviation used to determine sample size, the allowance for sampling risk might not have been adequately controlled.

7.34 If the allowance for sampling risk has not been adequately limited (for example, the sample was too small), the auditor may

    a. examine additional randomly selected sample items if the auditor determines that extending the sample is appropriate. The auditor may calculate the additional sample size using a revised estimate of the variation in the population such that the total number of sampling units in the additional sample combined with the original sample can be expected to adequately limit the allowance for sampling risk. Adding only a few additional items to the original sample is usually an ineffective procedure, and often the sample may need to be at least doubled to have a significant effect on the computed limit(s), but recomputing the required sample size to meet the test objectives provides specific guidance for expanding a sample.

---

[10] *Note:* This is not the case in this example. If the limit obtained from the sample was below $140,000, then this would be the case.

Classical Variables Sampling **121**

    b. perform additional substantive tests such as analytical procedures directed toward the same audit objective. The additional reliance on other tests would allow the auditor to accept a greater risk of incorrect acceptance for the sampling application. Recalculating the allowance for sampling risk with the greater risk of incorrect acceptance does not change the point estimate of the population, but it does move the ends of the range closer to the point estimate. In general, this approach may only be effective when differences between the desired and achieved results are small because other tests may not provide the quality of direct evidence regarding the population that a sample might provide.

7.35 Although the auditor selects a sample in such a way that it can be expected to be representative of the population, occasionally the sample might not be typical of the whole; thus, the sample results might not support acceptance of the population's recorded amounts. The auditor might have reason to believe that the sample is not representative of the population if, for example, other related audit evidence contradicts the sample evidence. In this situation, the auditor might suspect, among other possibilities, that the sample consists of items with small or large amounts or items with a rate of misstatement that are not representative of the population. It is important for the auditor considering such a judgment to recognize that the sample is expected to be representative only with respect to the occurrence rate or incidence of misstatements, not their nature. An unusual sample misstatement may be indicative of other unusual misstatements in the population. When the auditor concludes the sample may not be representative, he or she might examine additional sampling units or perform alternative procedures to determine whether the recorded amount of the population is misstated.[11]

7.36 In rare cases where significant related audit evidence outside the sample contradicts the sample evidence, the auditor might have a basis to suspect that the sample is not representative of the population. The general guidance of auditors with significant sampling experience is to "believe the sample," and only rarely is it appropriate to take out-of-the ordinary action when they encounter such a misstatement.

7.37 There will be times when there is no evidence that the sample is unrepresentative, but the auditor has not achieved the desired allowance for sampling risk (precision). In these situations, it is often appropriate to extend the sample or apply other audit procedures to achieve the desired allowance for sampling risk.

7.38 If the sample results do not support the recorded amount of the population and the auditor believes that the recorded amount may be misstated, he or she should consider the misstatement along with other audit evidence when evaluating whether the financial statements are materially misstated. As stated in paragraphs .A9–.A10 of AU-C section 450, *Evaluation of Misstatements Identified During the Audit* (AICPA, *Professional Standards*), the auditor may request that management examine the population to determine the cause and whether there are additional misstatements and, if appropriate, adjust the recorded amount. If the difference between the adjusted recorded amount and the far end of the range is less than the tolerable misstatement, the

---

[11] Paragraphs 4.101–.104 in chapter 4, "Nonstatistical and Statistical Audit Sampling for Substantive Tests of Details," provide further discussion of unusual sample results.

sample results would support the conclusion that the population, as adjusted, is not misstated by more than tolerable misstatement.

7.39 In addition to evaluating the frequency and amounts of monetary misstatements, the auditor should consider the qualitative aspects of misstatements. These considerations are discussed in chapter 4.

Classical Variables Sampling Case Study

7.40 ABC Co., a distributor of household products, is audited by Smith, Stein & Co., CPAs. Alexandra Stein of Smith, Stein & Co. decided to design a classical variables statistical sample to test the pricing of ABC Co.'s inventory as part of the audit of the company's June 30, 20XX financial statements. For the year ended June 30, 20XX, ABC Co.'s inventory, which consisted of approximately 2,700 different items, had a recorded amount of $3,207,892.50.

7.41 Stein decided that the results of her consideration and tests of ABC Co.'s internal control supported an assessed level of control risk at a moderate level for the assertion of valuation of inventories. She also decided that materiality for the entire audit was $90,000, performance materiality was $55,000, and that a misstatement of $45,000 or more from this sample of the inventory balance, when combined with possible misstatements from other tests, could result in the financial statements being materially misstated or fail to adequately allow for sampling risk, or both. In reaching these assessments, Stein considered and documented her rationale concerning the factors illustrated in table 4-3, "Factors to Consider in Setting Performance Materiality (PM) at the Engagement Level and Tolerable Misstatement (TM) at the Test Level."[12]

7.42 Stein chose a classical variables sampling approach because, on the basis of the prior year's audit, (*a*) she expected the account to contain both overstatements and understatements and expected some misstatements, and (*b*) the accounting records had been maintained on a computer. She had computer software to analyze the accounting records and assist her in designing and evaluating the sample.

7.43 Stein obtained assurance that inventory quantities were recorded properly by observing ABC Co.'s physical inventory as of June 30, 20XX, and applying cutoff procedures. She planned to perform some analytical procedures on the inventory account to obtain further assurance that both the quantities and pricing were reasonable. Although Stein expected to find some misstatements, she did not expect to find enough misstatements to use either a ratio or a difference sampling approach. Therefore, she decided to design a mean-per-unit statistical sample. If she found enough misstatements, she could evaluate the sample result using a difference or ratio approach.

7.44 The approximately 2,700 items of ABC Co.'s inventory balance had a wide range of recorded amounts, from approximately $20 to $7,500 per item. Stein decided to stratify the items constituting the balance to reduce the effect that variation in recorded amounts had on the determination of sample size. She identified 9 items whose recorded amounts each exceeded $4,500. Those items were examined 100 percent and were not to be included in the items subject to sampling.

---

[12] Had Stein considered all or most of the relevant factors when setting performance materiality, then tolerable misstatement might be the same or slightly less than performance materiality in this circumstance.

7.45 Using professional judgment, Stein decided that a 20 percent risk of incorrect acceptance (in other words, 80 percent confidence) was appropriate for this test because of the moderate assessed level of risk of material misstatement (including control risk), and the moderate reliance she intended to place on other planned substantive tests related to the assertion of valuation of the inventory account.[13] In calculating the sample size, Stein also decided to specify a 15 percent risk of incorrect rejection to provide a sample size that would be large enough to allow for some misstatement.

7.46 Because ABC Co.'s inventory records were maintained on a computer, Stein was able to use a computer program to assist her in stratifying the June 30, 20XX, inventory and in selecting an appropriate sample. The computer program divided the items subject to sampling into 10 strata and calculated an appropriate sample size for each stratum (see exhibit 7-2, "Inventory Sample Evaluation Report"). The overall sample size calculated by the program, based on the risk levels and tolerable misstatement specified by Stein, was 209 (see exhibit 7-2). The total sample size of 209 consisted of 200 items selected from the population subject to sampling and 9 items to be examined 100 percent due to their size and risk. Stein tested the pricing of the 209 inventory items and identified 6 misstatements: 5 in the sample of 200 and 1 overstatement in the 9 items examined 100 percent.[14]

7.47 Stein used another computer program to assist her in calculating the projected misstatement and the allowance for sampling risk for the sample. That program calculated a projected misstatement for each stratum and total factual and projected misstatement and allowance for sampling risk for the entire sample at the 20 percent risk of incorrect acceptance she had specified (see exhibit 7-2). The total factual and projected misstatement was $16,394.48 ($3,207,892.50 − $3,191,498.02).

7.48 Because the total factual and projected misstatement of $16,394.48 in the inventory balance ($14,394.48 projected from the population subject to sampling plus $2,000 of misstatement identified in the items examined 100 percent) plus a $21,222.11 allowance for sampling risk (see exhibit 7-2) was less than the $45,000 tolerable misstatement for the inventory balance, Stein concluded that the sample results supported ABC Co.'s recorded amount of inventory; however, she aggregated the projected misstatement from the sample with other factual and projected (likely) misstatements[15] when she evaluated whether the financial statements as a whole were materially misstated. She also brought the factual and projected (likely) misstatement to management's attention. Management did not make any adjustments except for the identified, factual misstatements. There were no zero or negative items in the population.

---

[13] A consideration of the audit risk relationships, illustrated in paragraphs 4.39–.42 of this guide, might also illustrate the appropriateness of the 80 percent assurance by noting that the risks of risk of material misstatement (RMM) (after testing controls to, for example, limit RMM to 50 percent), substantive details tests (at 20 percent risk), and analytical procedures (which were deemed 50 percent effective in detecting tolerable misstatement, for example) result in a low risk (for example, 0.50 RMM × 0.20 detail tests × 0.50 analytical = 0.05 risk).

[14] Stein's firm does not require (and her software does not compute) a minimum sample size per stratum. She believes the strata sizes of 17–24 are adequate for this test.

[15] Any judgmental differences should also be aggregated but are not discussed in this guide.

Exhibit 7-1

Inventory Sample Size Report
ABC Co.
June 30, 20XX

Stratum Number	Stratum Low Range	Stratum High Range	Total Items in Stratum	Standard Deviation	Sample Size
1	0	236	420	62.38	21
2	237	450	409	65.06	21
3	451	663	390	62.23	19
4	664	911	356	68.65	19
5	912	1,260	308	101.21	24
6	1,261	1,698	187	123.70	18
7	1,699	2,441	127	212.92	21
8	2,442	3,116	144	181.52	21
9	3,117	3,555	205	113.52	19
10	3,556	4,500	148	145.71	17
100%	4,500	—	9	—	9

Recorded amount of population	$3,207,892.50
Total sampling units in population	2,695
Total sample size	209

The sample was calculated based on the following specifications:

Tolerable misstatement	45,000
Risk of incorrect acceptance	0.20
Risk of incorrect rejection	0.15
Lower 100 percent cutoff	0
Upper 100 percent cutoff	4,500

AAG-SAM 7.48

Exhibit 7-2

Inventory Sample Evaluation Report
ABC Co.
June 30, 20XX

Misstatements Located in Audit	Recorded Amount	Audit Amount
1	$1,250.00	$350.00
2	200.00	360.00
3	600.00	240.00
4	510.00	650.00
5	320.00	319.00
6	7,550.00	5,550.00
TOTAL	$10,430.00	$7,469.00

Estimated total amount	3,191,498.02
Allowance for sampling risk	21,222.11
Sampling units in population	2,695
Sample size	209
Tolerable misstatement	45,000.00
Risk of incorrect acceptance	0.20
Risk of incorrect rejection	0.15

Variables test evaluation:

Recorded amount of $3,207,892.50 can be accepted as not misstated by more than a tolerable amount given the tolerable misstatement originally specified if the risk of incorrect acceptance of 0.20 for this test remains appropriate after considering the results of other auditing procedures.

Appendix A
Attributes Statistical Sampling Tables

A.1 Four tables appear at the end of this appendix to assist the auditor in planning and evaluating a statistical sample of a fixed size for a test of controls.[1] They are as follows:

- Table A-1, "Statistical Sample Sizes for Tests of Controls—5 Percent Risk of Overreliance"[2]
- Table A-2, "Statistical Sample Sizes for Tests of Controls—10 Percent Risk of Overreliance"
- Table A-3, "Statistical Sampling Results Evaluation Table for Tests of Controls—Upper Limits at 5 Percent Risk of Overreliance"
- Table A-4, "Statistical Sampling Results Evaluation Table for Tests of Controls—Upper Limits at 10 Percent Risk of Overreliance"

Using the Tables

A.2 Chapter 3, "Nonstatistical and Statistical Audit Sampling in Tests of Controls," discusses the factors that the auditor needs to consider when planning an audit sampling application for a test of controls. For statistical sampling, the auditor needs to specify explicitly (*a*) an acceptable level of the risk of overreliance, (*b*) the tolerable rate of deviation, and (*c*) the expected population deviation rate. This appendix includes tables for 5 percent and 10 percent levels of risk of assessing controls as effective when they are not (overreliance). Either a table in another reference on statistical sampling or a computer program is necessary if the auditor desires another level of risk of overreliance.[3]

A.3 The auditor selects the table for the acceptable level of risk and then reads down the expected population deviation rate column to find the appropriate rate. Next, the auditor locates the column corresponding to the tolerable rate of deviation. The appropriate sample size is shown where the two factors meet.

A.4 In some circumstances, tables A-1 and A-2 may be used to evaluate the sample results. The parenthetical number shown next to each sample size is the expected number of deviations planned for in the sample. The expected number of deviations is the expected population deviation rate multiplied by the sample size. If the auditor finds that number of deviations or fewer in the

---

[1] Auditors using a sequential sampling plan should not use these tables for designing or evaluating the sample application. See the discussion of sequential sampling in appendix B, "Sequential Sampling for Tests of Controls."

[2] The risk that the tolerable rate of deviation is exceeded by the actual rate of deviation in the population (also, the risk that the controls will be assessed as more effective than they actually are).

[3] Other methods in this guide may also provide acceptable approximations of attribute sample sizes; for example, the discussion in paragraph 4.72 and table 4-6, "Confidence (Reliability) Factors," of chapter 4, "Nonstatistical and Statistical Audit Sampling for Substantive Tests of Details," of this guide.

sample, he or she can conclude (at a minimum) that at the desired risk, the projected deviation rate for the population, plus an allowance for sampling risk, is not more than the tolerable rate. In these circumstances, the auditor need not use table A-3 or A-4 to evaluate the sample results.

A.5 If more than the expected number of deviations are found in the sample, the auditor cannot conclude at the desired risk of overreliance that the population deviation rate is less than the tolerable rate. Accordingly, the test would not support his or her planned assessment of control risk; however, the sample might support some lesser assessment (for example, at a higher level of risk or a greater level of tolerable deviation rate).

A.6 If the number of deviations found in the sample is not the expected number of deviations shown in the parentheses in tables A-1 or A-2, and the auditor wishes to calculate the maximum (for example, upper statistical limit) deviation rate in the population, he or she can evaluate the sample results using either table A-3, for a 5 percent acceptable risk of overreliance, or table A-4, for a 10 percent acceptable risk of overreliance. Space limitations do not allow tables A-3 and A-4 to include evaluations for all possible sample sizes or for all possible numbers of deviations found. If the auditor is evaluating sample results for a sample size or number of deviations not shown in these tables, he or she may be able to use either a table in another reference on statistical sampling or a computer program. Alternatively, the auditor might interpolate between sample sizes shown in these tables. Any error due to interpolation is generally not significant to the auditor's evaluation. If the auditor wishes to be conservative, he or she can use the next smaller sample size shown in the table to evaluate the number of deviations found in the sample.

A.7 The auditor uses the table applicable to the acceptable level of risk of overreliance and then reads down the sample-size column to find the appropriate sample size. Next, the auditor locates the column corresponding to the number of deviations found in the sample. The projection of the sample results to the population plus an allowance for sampling risk (that is, the maximum population deviation rate) is shown where the two factors meet. If this maximum population deviation rate is less than the tolerable rate, the test supports the planned assessment of control risk.

Applying Nonstatistical Sampling for Tests of Controls

A.8 The auditor, using nonstatistical sampling for tests of controls, uses his or her professional judgment to consider the factors described in chapter 3 in determining sample sizes. The relative effect of each factor on the appropriate nonstatistical sample size is illustrated in chapter 3 and is summarized in exhibit A-1.

Exhibit A-1

Determining Sample Sizes

Factor	General Effect on Sample Size
Tolerable rate increase (decrease)	Smaller (larger)
Risk of overreliance increase (decrease)	Smaller (larger)
Expected population deviation rate increase (decrease)	Larger (smaller)
Population size	Virtually no effect[1]

[1] Unless the population is very small.

A.9 Neither paragraph .A14 of AU-C section 530, *Audit Sampling* (AICPA, *Professional Standards*), nor this guide requires the auditor to compute the sample size for a nonstatistical sampling application with a corresponding sample size calculated using statistical theory; however, in applying informed professional judgment to determine an appropriate nonstatistical sample size for a test of controls, an auditor might find it helpful to be familiar with the tables in this appendix. The auditor using these tables as an aid in understanding relative sample sizes for tests of controls will need to apply professional judgment in specifying the risk levels and expected population deviation rates in relation to sample sizes. For example, an auditor designing a nonstatistical sampling application to test compliance with a prescribed control procedure might have assessed the tolerable rate as 8 percent. If the auditor were to consider selecting a sample size of 60, these tables would imply that at approximately a 5 percent risk level, the auditor expected no more than approximately 1.5 percent of the items in the population to be deviations from the prescribed control procedure. These tables also would imply that at approximately a 10 percent risk level, the auditor expected no more than approximately 3 percent of the items in the population to be deviations.

A.10 These tables were designed for attributes sampling (for example, tests of controls) where a deviation is or is not present in each individual sample item. They may be used for determining a monetary unit sampling sample size when expected misstatement is zero or where the expected taint of any misstatement found is assumed to be a 100 percent taint (a conservative planning assumption).

Basis for Tables A-1–A-4

A.11 The tables were computed using the binomial distribution and assume a large population. Sample sizes in tables A-1 and A-2 were rounded upward (for example, 51.01 becomes 52). Evaluations in tables A-3 and A-4 were rounded upward (5.01 percent becomes 5.1 percent). The expected number of deviations in tables A-1 and A-2 was rounded upward (0.2 deviations becomes 1 deviation) and the sample size computed is based on the rounded number of deviations expected.

AAG-SAM APP A

Table A-1
Statistical Sample Sizes for Tests of Controls—5 Percent Risk of Overreliance
(with number of expected errors in parentheses)

Expected Deviation Rate	Tolerable Deviation Rate									
	2%	3%	4%	5%	6%	7%	8%	9%	10%	15%	20%
0.00%	149 (0)	99 (0)	74 (0)	59 (0)	49 (0)	42 (0)	36 (0)	32 (0)	29 (0)	19 (0)	14 (0)
0.25%	236 (1)	157 (1)	117 (1)	93 (1)	78 (1)	66 (1)	58 (1)	51 (1)	46 (1)	30 (1)	22 (1)
0.50%	313 (2)	157 (1)	117 (1)	93 (1)	78 (1)	66 (1)	58 (1)	51 (1)	46 (1)	30 (1)	22 (1)
0.75%	386 (3)	208 (2)	117 (1)	93 (1)	78 (1)	66 (1)	58 (1)	51 (1)	46 (1)	30 (1)	22 (1)
1.00%	590 (6)	257 (3)	156 (2)	93 (1)	78 (1)	66 (1)	58 (1)	51 (1)	46 (1)	30 (1)	22 (1)
1.25%	1,030 (13)	303 (4)	156 (2)	124 (2)	78 (1)	66 (1)	58 (1)	51 (1)	46 (1)	30 (1)	22 (1)
1.50%		392 (6)	192 (3)	124 (2)	103 (2)	66 (1)	58 (1)	51 (1)	46 (1)	30 (1)	22 (1)
1.75%		562 (10)	227 (4)	153 (3)	103 (2)	88 (2)	77 (2)	51 (1)	46 (1)	30 (1)	22 (1)
2.00%		846 (17)	294 (6)	181 (4)	127 (3)	88 (2)	77 (2)	68 (2)	46 (1)	30 (1)	22 (1)
2.25%		1,466 (33)	390 (9)	208 (5)	127 (3)	88 (2)	77 (2)	68 (2)	61 (2)	30 (1)	22 (1)
2.50%			513 (13)	234 (6)	150 (4)	109 (3)	77 (2)	68 (2)	61 (2)	30 (1)	22 (1)
2.75%			722 (20)	286 (8)	173 (5)	109 (3)	95 (3)	68 (2)	61 (2)	30 (1)	22 (1)
3.00%			1,098 (33)	361 (11)	195 (6)	129 (4)	95 (3)	84 (3)	61 (2)	30 (1)	22 (1)
3.25%			1,936 (63)	458 (15)	238 (8)	148 (5)	112 (4)	84 (3)	61 (2)	30 (1)	22 (1)
3.50%				624 (22)	280 (10)	167 (6)	112 (4)	84 (3)	76 (3)	40 (2)	22 (1)
3.75%				877 (33)	341 (13)	185 (7)	129 (5)	100 (4)	76 (3)	40 (2)	22 (1)
4.00%				1,348 (54)	421 (17)	221 (9)	146 (6)	100 (4)	89 (4)	40 (2)	22 (1)
5.00%					1,580 (79)	478 (24)	240 (12)	158 (8)	116 (6)	40 (2)	30 (2)
6.00%						1,832 (110)	532 (32)	266 (16)	179 (11)	50 (3)	30 (2)
7.00%								585 (41)	298 (21)	68 (5)	37 (3)
8.00%									649 (52)	85 (7)	37 (3)
9.00%										110 (10)	44 (4)
10.00%										150 (15)	50 (5)
12.50%										576 (72)	88 (11)
15.00%											193 (29)
17.50%											720 (126)

Note: Sample sizes over 2,000 items not shown. This table assumes a large population.

Table A-2

Statistical Sample Sizes for Tests of Controls—10 Percent Risk of Overreliance (with number of expected errors in parentheses)

Expected Deviation Rate	Tolerable Deviation Rate									
	2%	3%	4%	5%	6%	7%	8%	9%	10%	15%	20%
0.00%	114 (0)	76 (0)	57 (0)	45 (0)	38 (0)	32 (0)	28 (0)	25 (0)	22 (0)	15 (0)	11 (0)
0.25%	194 (1)	129 (1)	96 (1)	77 (1)	64 (1)	55 (1)	48 (1)	42 (1)	38 (1)	25 (1)	18 (1)
0.50%	194 (1)	129 (1)	96 (1)	77 (1)	64 (1)	55 (1)	48 (1)	42 (1)	38 (1)	25 (1)	18 (1)
0.75%	265 (2)	129 (1)	96 (1)	77 (1)	64 (1)	55 (1)	48 (1)	42 (1)	38 (1)	25 (1)	18 (1)
1.00%	398 (4)	176 (2)	96 (1)	77 (1)	64 (1)	55 (1)	48 (1)	42 (1)	38 (1)	25 (1)	18 (1)
1.25%	708 (9)	221 (3)	132 (2)	77 (1)	64 (1)	55 (1)	48 (1)	42 (1)	38 (1)	25 (1)	18 (1)
1.50%	1,463 (22)	265 (4)	132 (2)	105 (2)	64 (1)	55 (1)	48 (1)	42 (1)	38 (1)	25 (1)	18 (1)
1.75%		390 (7)	166 (3)	105 (2)	88 (2)	55 (1)	48 (1)	42 (1)	38 (1)	25 (1)	18 (1)
2.00%		590 (12)	198 (4)	132 (3)	88 (2)	75 (2)	48 (1)	42 (1)	38 (1)	25 (1)	18 (1)
2.25%		974 (22)	262 (6)	132 (3)	88 (2)	75 (2)	65 (2)	42 (1)	38 (1)	25 (1)	18 (1)
2.50%			353 (9)	158 (4)	110 (3)	75 (2)	65 (2)	58 (2)	38 (1)	25 (1)	18 (1)
2.75%			471 (13)	209 (6)	132 (4)	94 (3)	65 (2)	58 (2)	52 (2)	25 (1)	18 (1)
3.00%			730 (22)	258 (8)	132 (4)	94 (3)	65 (2)	58 (2)	52 (2)	25 (1)	18 (1)
3.25%			1,258 (41)	306 (10)	153 (5)	113 (4)	82 (3)	58 (2)	52 (2)	25 (1)	18 (1)
3.50%				400 (14)	194 (7)	113 (4)	82 (3)	73 (3)	52 (2)	25 (1)	18 (1)
3.75%				583 (22)	235 (9)	131 (5)	98 (4)	73 (3)	52 (2)	25 (1)	18 (1)
4.00%				873 (35)	274 (11)	149 (6)	98 (4)	73 (3)	65 (3)	25 (1)	18 (1)
5.00%					1,019 (51)	318 (16)	160 (8)	115 (6)	78 (4)	34 (2)	18 (1)
6.00%						1,150 (69)	349 (21)	182 (11)	116 (7)	43 (3)	25 (2)
7.00%							1,300 (91)	385 (27)	199 (14)	52 (4)	25 (2)
8.00%								1,437 (115)	424 (34)	60 (5)	25 (2)
9.00%									1,577 (142)	77 (7)	32 (3)
10.00%										100 (10)	38 (4)
12.50%										368 (46)	63 (8)
15.00%											126 (19)
17.50%											457 (80)

Note: Sample sizes over 2,000 items not shown. This table assumes a large population.

Table A-3

Statistical Sampling Results Evaluation Table for Tests of Controls—Upper Limits at 5 Percent Risk of Overreliance

Sample Size	\multicolumn{11}{c}{Actual Number of Deviations Found}

Sample Size	0	1	2	3	4	5	6	7	8	9	10
20	14.0	21.7	28.3	34.4	40.2	45.6	50.8	55.9	60.7	65.4	69.9
25	11.3	17.7	23.2	28.2	33.0	37.6	42.0	46.3	50.4	54.4	58.4
30	9.6	14.9	19.6	23.9	28.0	31.9	35.8	39.4	43.0	46.6	50.0
35	8.3	12.9	17.0	20.7	24.3	27.8	31.1	34.4	37.5	40.6	43.7
40	7.3	11.4	15.0	18.3	21.5	24.6	27.5	30.4	33.3	36.0	38.8
45	6.5	10.2	13.4	16.4	19.2	22.0	24.7	27.3	29.8	32.4	34.8
50	5.9	9.2	12.1	14.8	17.4	19.9	22.4	24.7	27.1	29.4	31.6
55	5.4	8.4	11.1	13.5	15.9	18.2	20.5	22.6	24.8	26.9	28.9
60	4.9	7.7	10.2	12.5	14.7	16.8	18.8	20.8	22.8	24.8	26.7
65	4.6	7.1	9.4	11.5	13.6	15.5	17.5	19.3	21.2	23.0	24.7
70	4.2	6.6	8.8	10.8	12.7	14.5	16.3	18.0	19.7	21.4	23.1
75	4.0	6.2	8.2	10.1	11.8	13.6	15.2	16.9	18.5	20.1	21.6
80	3.7	5.8	7.7	9.5	11.1	12.7	14.3	15.9	17.4	18.9	20.3
90	3.3	5.2	6.9	8.4	9.9	11.4	12.8	14.2	15.5	16.9	18.2
100	3.0	4.7	6.2	7.6	9.0	10.3	11.5	12.8	14.0	15.2	16.4
125	2.4	3.8	5.0	6.1	7.2	8.3	9.3	10.3	11.3	12.3	13.2
150	2.0	3.2	4.2	5.1	6.0	6.9	7.8	8.6	9.5	10.3	11.1
200	1.5	2.4	3.2	3.9	4.6	5.2	5.9	6.5	7.2	7.8	8.4
300	1.0	1.6	2.1	2.6	3.1	3.5	4.0	4.4	4.8	5.2	5.6
400	0.8	1.2	1.6	2.0	2.3	2.7	3.0	3.3	3.6	3.9	4.3
500	0.6	1.0	1.3	1.6	1.9	2.1	2.4	2.7	2.9	3.2	3.4

Note: This table presents upper limits (body of table) as percentages. This table assumes a large population

Table A-4

Statistical Sampling Results Evaluation Table for Tests of Controls—Upper Limits at 10 Percent Risk of Overreliance

Sample Size	Actual Number of Deviations Found										
	0	1	2	3	4	5	6	7	8	9	10
20	10.9	18.1	24.5	30.5	36.1	41.5	46.8	51.9	56.8	61.6	66.2
25	8.8	14.7	20.0	24.9	29.5	34.0	38.4	42.6	46.8	50.8	54.8
30	7.4	12.4	16.8	21.0	24.9	28.8	32.5	36.2	39.7	43.2	46.7
35	6.4	10.7	14.5	18.2	21.6	24.9	28.2	31.4	34.5	37.6	40.6
40	5.6	9.4	12.8	16.0	19.0	22.0	24.9	27.7	30.5	33.2	35.9
45	5.0	8.4	11.4	14.3	17.0	19.7	22.3	24.8	27.3	29.8	32.2
50	4.6	7.6	10.3	12.9	15.4	17.8	20.2	22.5	24.7	27.0	29.2
55	4.2	6.9	9.4	11.8	14.1	16.3	18.4	20.5	22.6	24.6	26.7
60	3.8	6.4	8.7	10.8	12.9	15.0	16.9	18.9	20.8	22.7	24.6
65	3.5	5.9	8.0	10.0	12.0	13.9	15.7	17.5	19.3	21.0	22.8
70	3.3	5.5	7.5	9.3	11.1	12.9	14.6	16.3	18.0	19.6	21.2
75	3.1	5.1	7.0	8.7	10.4	12.1	13.7	15.2	16.8	18.3	19.8
80	2.9	4.8	6.6	8.2	9.8	11.3	12.8	14.3	15.8	17.2	18.7
90	2.6	4.3	5.9	7.3	8.7	10.1	11.5	12.8	14.1	15.4	16.7
100	2.3	3.9	5.3	6.6	7.9	9.1	10.3	11.5	12.7	13.9	15.0
125	1.9	3.1	4.3	5.3	6.3	7.3	8.3	9.3	10.2	11.2	12.1
150	1.6	2.6	3.6	4.4	5.3	6.1	7.0	7.8	8.6	9.4	10.1
200	1.2	2.0	2.7	3.4	4.0	4.6	5.3	5.9	6.5	7.1	7.6
300	0.8	1.3	1.8	2.3	2.7	3.1	3.5	3.9	4.3	4.7	5.1
400	0.6	1.0	1.4	1.7	2.0	2.4	2.7	3.0	3.3	3.6	3.9
500	0.5	0.8	1.1	1.4	1.6	1.9	2.1	2.4	2.6	2.9	3.1

Note: This table presents upper limits (body of table) as percentages. This table assumes a large population

Appendix B

Sequential Sampling for Tests of Controls

B.1 The auditor designs samples for tests of controls using either a fixed sampling plan or a sequential sampling plan.[1] Under a fixed sampling plan, the auditor examines a single sample of a specified size; under a sequential sampling plan, the sample is selected in several steps, with each step conditional on the results of the previous steps. The decision to use a fixed or a sequential sampling plan depends on which plan the auditor believes is more efficient in the circumstances.

B.2 In planning a fixed sampling application, the auditor considers that if the deviation rate in the sample exceeds the specified expected population deviation rate, the sample results would suggest that the estimated population deviation rate plus an allowance for sampling risk exceeds the tolerable rate of deviation. In that case, the sample results would not support the auditor's planned assessed level of control risk. These results might be obtained even though the actual population deviation rate would support the auditor's planned assessment because the sample size is too small to limit adequately the allowance for sampling risk. Additionally, the deviation rate observed in the sample may be higher than expected because the sample is not representative of the true deviation rate in the population.

B.3 Consequently, in a fixed sampling application, the sample either passes or fails and in a statistical application is not extended to mitigate the effect of unexpected deviations that may appear in a sample. The auditor can use a sequential sampling plan to help overcome this limitation of a fixed sampling plan.

B.4 A sequential sample generally consists of two to four groups of sampling units. The auditor determines the sizes of the individual groups of sampling units based on the specified risk of overreliance, the tolerable rate of deviation, and the expected population deviation rate. The auditor generally uses a computer program or specially designed tables for sequential sampling plans to assist in determining the appropriate size for each group of sampling units. While a number of texts and publications provide a number of plans, a sampling specialist is often consulted when developing a custom plan, as valid sequential plans are not developed directly from conventional single stage tables and software. In a valid sequential plan, the plan includes a consideration that the decision to move to a second or subsequent stage brings a risk that the next stage of the sample will reveal fewer deviations than would be representative from the population, thereby increasing the overall risk of incorrect acceptance.

B.5 In a sequential sample, the auditor examines the first group of sampling units and, on the basis of the results, decides whether to (*a*) accept the assessed level of control risk as planned, without examining additional sampling units, (*b*) stop sampling because the planned confidence and tolerable rate of deviation cannot be achieved as too many deviations were found, thus

---

[1] More discussion of designing a sequential sample can be found in Donald Roberts, *Statistical Auditing* (New York: AICPA, 1978): 57–60.

increasing the assessed level of control risk, or (c) examine additional sampling units because sufficient information to determine whether the planned assessed level of control risk is supported has not yet been obtained.

Example of a Sequential Sampling Plan

B.6 Table B-1, "Four Step Sequential Sampling Plan," illustrates the number of sampling units for each group in a four step sequential sampling plan, assuming a 5 percent tolerable rate of deviation, a 10 percent risk of assessing control risk too low, a 5 percent risk of overreliance, and a 0.5 percent population deviation rate related to assessing control risk too high. This plan requires the increments between each step to be the same number (after the first step of 50, each additional step is 51).

Table B-1

Four Step Sequential Sampling Plan

			Accumulated Deviation		
Group	Number of Sampling Units	Accumulated Sampling Units	Accept Planned Assessed Level	Sample More	Increase Planned Assessed Level
1	50	50	0	1–3	4
2	51	101	1	2–3	4
3	51	152	2	3	4
4	51	203	3	N/A	4

B.7 If the auditor finds 4 deviations at any time in this example, the examination of sampling units stops and the assessed level of control risk is increased beyond that which was planned. If no deviations are found in the first group of 50 sampling units, the auditor concludes that the sample supports the planned assessed level without examining more sampling units. If 1, 2, or 3 deviations exist in the first group of sampling units, the auditor examines additional sampling units in the next group(s). The auditor continues to examine sampling units in succeeding groups until the sample results either support or do not support the planned assessed level. For example, if 3 deviations exist in the first group, the next 3 groups of sampling units are examined without finding additional deviations to support the planned assessed level of control risk.

B.8 To achieve statistically valid conclusions, the auditor follows the rules of the plan. Thus, consideration is given at the outset of the number of stages that are to be used in the plan. The four step plan previously illustrated may cause the auditor to test more than 200 instances of a single control, depending on the outcome of each stage. In the end, the auditor may still have to reject the control as ineffective when additional deviations are found. Thus, auditors consider the cost-benefit (for example, considering the effect on substantive testing and the effectiveness of controls versus substantive assurance) of extensive control testing and seek to limit the extent of control testing by limiting the sequential plan to two or three stages (see table B-2).

Comparison of Sequential Sample Sizes With Fixed Sample Sizes

B.9 Sample sizes under fixed sampling plans are larger, on the average, than those under sequential sampling plans if the auditor overstates the expected population deviation rate. For example, if the actual population deviation rate is 0.5 percent, the four step sequential sampling plan illustrated in table B-1 would generally require the auditor to examine fewer sampling units to support the planned assessed level than a fixed sampling plan would require; however, if the auditor finds one deviation in the first group of sample items, the auditor will test more items under a sequential plan, and may even have to move on to additional stages depending on the stage when the deviations are found.

B.10 Under a fixed sampling plan, a sample size of 77 is sufficient to support the planned assessed level when the population deviation rate is 0.5 percent (see table A-2 in appendix A, "Attributes Statistical Sampling Tables"). Under the sequential sampling plan, the auditor examines 50, 101, 152, or 203 items; however, in addition to the cost-benefit of applying sequential sampling in a specific instance, the auditor considers the long-run average sample size indicated to meet his or her objectives. For example, if the true population deviation rate is 0.5 percent, the auditor may need to examine an average of 65 sampling units under the four step sequential sampling plan as compared with 77 sampling units under the fixed sampling plan.

B.11 A sequential sampling plan provides an opportunity to minimize sampling in populations with a low deviation rate; however, an auditor might find that the audit effort of examining the total number of sampling units for all four steps of a sequential sampling plan would exceed the reduction of substantive testing that could be achieved by performing tests of controls. The auditor may stop testing at any time and assume the control is not effective at the level of sample assurance desired, and plan other (for example, substantive) tests accordingly.

B.12 If the auditor believes it would not be practical to examine the total number of sampling units for all steps of a four step sequential sampling plan, a sequential sampling plan with fewer than four steps could be designed. For example, some auditors find it practical to design two step sequential sampling plans.

B.13 The following two stage plan[2] is designed at a 10 percent risk of overreliance. For the following plan, the decision rule allows the auditor to stop at the end of the first sample if no deviations are found. If only one deviation is encountered during the first stage sample, the auditor extends the sample to the second stage. If a second deviation is found either in the first or second stage, the auditor will not be able to achieve the desired sample result even if no additional deviations are found.

---

[2] See Vincent M. O'Reilly et al., *Montgomery's Auditing, 12th Edition* (Wiley, 1999): 16:47. The table was computed with a focus on minimizing the first stage sample size.

Table B-2

Tolerable Rate of Deviation	1st Sample	2nd Sample
10%	23	29
8%	30	30
5%	51	39
3%	89	56
2%	133	87

B.14 Sequential sampling plans are generally designed for statistical sampling applications; however, they might also be used in a nonstatistical sampling application.

Appendix C

Monetary Unit Sampling Tables

C.1 *Note:* For identical risks of incorrect acceptance,[1] sample sizes determined by table 4-5, "Illustrative Sample Sizes" (table C-1, "Monetary Unit Sample Size Determination Tables") and table C-2, "Confidence Factors for Monetary Unit Sample Size Design," will be the same.

---

[1] The risk that the auditor will conclude that a misstatement greater than tolerable misstatement does not exist when it does.

Table C-1

Monetary Unit Sample Size Determination Tables

Risk of Incorrect Acceptance	Ratio of Expected to Tolerable Misstatement	Tolerable Misstatement as a Percentage of Population											Expected Sum of Taints
		50%	30%	10%	8%	6%	5%	4%	3%	2%	1%	0.50%	
5%	—	6	10	30	38	50	60	75	100	150	300	600	—
5%	0.10	8	13	37	46	62	74	92	123	184	368	736	0.37
5%	0.20	10	16	47	58	78	93	116	155	232	463	925	0.93
5%	0.30	12	20	60	75	100	120	150	200	300	600	1,199	1.80
5%	0.40	17	27	81	102	135	162	203	270	405	809	1,618	3.24
5%	0.50	24	39	116	145	193	231	289	385	577	1,154	2,308	5.77
10%	—	5	8	24	29	39	47	58	77	116	231	461	—
10%	0.20	7	12	35	43	57	69	86	114	171	341	682	0.69
10%	0.30	9	15	44	55	73	87	109	145	217	433	866	1.30
10%	0.40	12	20	58	72	96	115	143	191	286	572	1,144	2.29
10%	0.50	16	27	80	100	134	160	200	267	400	799	1,597	4.00
15%	—	4	7	19	24	32	38	48	64	95	190	380	—
15%	0.20	6	10	28	35	46	55	69	91	137	273	545	0.55
15%	0.30	7	12	35	43	57	69	86	114	171	341	681	1.03
15%	0.40	9	15	45	56	74	89	111	148	221	442	883	1.77
15%	0.50	13	21	61	76	101	121	151	202	302	604	1,208	3.02

Monetary Unit Sampling Tables

Risk of Incorrect Acceptance	Ratio of Expected to Tolerable Misstatement	50%	30%	10%	8%	6%	5%	4%	3%	2%	1%	0.50%	Expected Sum of Taints
20%	—	4	6	17	21	27	33	41	54	81	161	322	—
20%	0.20	5	8	23	29	38	46	57	76	113	226	451	0.46
20%	0.30	6	10	28	35	47	56	70	93	139	277	554	0.84
20%	0.40	8	12	36	45	59	71	89	118	177	354	707	1.42
20%	0.50	10	16	48	60	80	95	119	159	238	475	949	2.38
25%	—	3	5	14	18	24	28	35	47	70	139	278	—
25%	0.20	4	7	19	24	32	38	48	64	95	190	380	0.38
25%	0.30	5	8	23	29	39	46	58	77	115	230	460	0.69
25%	0.40	6	10	29	37	49	58	73	97	145	289	578	1.16
25%	0.50	8	13	38	48	64	76	95	127	190	380	760	1.90
30%	—	3	5	13	16	21	25	31	41	61	121	241	—
30%	0.20	4	6	17	21	27	33	41	54	81	162	323	0.33
30%	0.40	5	8	24	30	40	48	60	80	120	239	477	0.96
30%	0.60	9	15	43	54	71	85	107	142	213	425	850	2.55
35%	—	3	4	11	14	18	21	27	35	53	105	210	—
35%	0.20	3	5	14	18	23	28	35	46	69	138	276	0.28
35%	0.40	4	7	20	25	34	40	50	67	100	199	397	0.80
35%	0.60	7	12	34	43	57	68	85	113	169	338	676	2.03
50%	—	2	3	7	9	12	14	18	24	35	70	139	—
50%	0.20	2	3	9	11	15	18	22	29	44	87	173	0.18
50%	0.40	3	4	12	15	19	23	29	38	57	114	228	0.46
50%	0.60	4	6	17	22	29	34	43	57	85	170	340	1.02

Tolerable Misstatement as a Percentage of Population

AAG-SAM APP C

C.2 As discussed in chapter 4, "Nonstatistical and Statistical Audit Sampling for Substantive Tests of Details," and chapter 6, "Monetary Unit Sampling," to determine sample size using table C-1 (also known as table 4-5), the auditor determines risk of incorrect acceptance, tolerable misstatement (as a percent of the population dollars), and expected misstatement (as a percentage of tolerable misstatement). Using these factors, the auditor finds the sample size in table 4-5. For example, if risk of incorrect acceptance is 10 percent, tolerable misstatement is 5 percent of the population dollars, and expected misstatement is 20 percent of tolerable misstatement (1 percent of the population dollars), the auditor identifies a sample size of 69.

C.3 For this sample size, the far right column of table 4-5 indicates that the sum of expected taints is 0.69.[2] The concept of taints comes from monetary unit sampling (MUS) and is discussed further in chapter 6. In performing the sample, the auditor may find complete and partial misstatements. A complete misstatement means the item has an audited amount of zero (for example, an account receivable of $1,000 that should be zero). An example of a partial misstatement is a $1,000 balance that should be $900 (this is a 10 percent partial misstatement or a 10 percent tainting). If the auditor found both previous two examples (one complete misstatement and one 10 percent tainting) the sum of the taints would be 1.10.

C.4 In the preceding example, if the auditor finds misstatements whose tainting percentages total to less than 0.69, he or she will be able to conclude at the stated risk of incorrect acceptance that it is unlikely that the population is misstated by more than 5 percent. If the auditor finds misstatements whose tainting percentages exceed 0.69, the auditor will not be able to conclude that the population is not misstated by more than 5 percent.

C.5 This table was based on the Poisson distribution, with sample sizes rounded to the next largest whole number.

---

[2] The sum of the expected tainting percentage was calculated by multiplying the sample size by the expected misstatements as a percentage of the population dollars. In the preceding case, the sample size was 69 and the expected misstatement was 1 percent of the population dollars thus the expected tainting was 0.69.

Table C-2

Confidence Factors for Monetary Unit Sample Size Design

Ratio of Expected to Tolerable Misstatement	Risk of Incorrect Acceptance								
	5%	10%	15%	20%	25%	30%	35%	37%	50%
0.00	3.00	2.31	1.90	1.61	1.39	1.21	1.05	1.00	0.70
0.05	3.31	2.52	2.06	1.74	1.49	1.29	1.12	1.06	0.73
0.10	3.68	2.77	2.25	1.89	1.61	1.39	1.20	1.13	0.77
0.15	4.11	3.07	2.47	2.06	1.74	1.49	1.28	1.21	0.82
0.20	4.63	3.41	2.73	2.26	1.90	1.62	1.38	1.30	0.87
0.25	5.24	3.83	3.04	2.49	2.09	1.76	1.50	1.41	0.92
0.30	6.00	4.33	3.41	2.77	2.30	1.93	1.63	1.53	0.99
0.35	6.92	4.95	3.86	3.12	2.57	2.14	1.79	1.67	1.06
0.40	8.09	5.72	4.42	3.54	2.89	2.39	1.99	1.85	1.14
0.45	9.59	6.71	5.13	4.07	3.29	2.70	2.22	2.06	1.25
0.50	11.54	7.99	6.04	4.75	3.80	3.08	2.51	2.32	1.37
0.55	14.18	9.70	7.26	5.64	4.47	3.58	2.89	2.65	1.52
0.60	17.85	12.07	8.93	6.86	5.37	4.25	3.38	3.09	1.70

Note: The basis for this table is the Poisson distribution. The 37 percent risk of incorrect acceptance column is provided for the convenience of those auditors that used previous MUS sampling formula guidance in developing policies and procedures.

AAG-SAM APP C

Table C-3
Monetary Unit Sampling—Confidence Factors for Sample Evaluation

Number of Overstatement Misstatements	Risk of Incorrect Acceptance								
	5%	10%	15%	20%	25%	30%	35%	37%	50%
0	3.00	2.31	1.90	1.61	1.39	1.21	1.05	1.00	0.70
1	4.75	3.89	3.38	3.00	2.70	2.44	2.22	2.14	1.68
2	6.30	5.33	4.73	4.28	3.93	3.62	3.35	3.25	2.68
3	7.76	6.69	6.02	5.52	5.11	4.77	4.46	4.35	3.68
4	9.16	8.00	7.27	6.73	6.28	5.90	5.55	5.43	4.68
5	10.52	9.28	8.50	7.91	7.43	7.01	6.64	6.50	5.68
6	11.85	10.54	9.71	9.08	8.56	8.12	7.72	7.57	6.67
7	13.15	11.78	10.90	10.24	9.69	9.21	8.79	8.63	7.67
8	14.44	13.00	12.08	11.38	10.81	10.31	9.85	9.68	8.67
9	15.71	14.21	13.25	12.52	11.92	11.39	10.92	10.74	9.67
10	16.97	15.41	14.42	13.66	13.02	12.47	11.98	11.79	10.67
11	18.21	16.60	15.57	14.78	14.13	13.55	13.04	12.84	11.67
12	19.45	17.79	16.72	15.90	15.22	14.63	14.09	13.89	12.67
13	20.67	18.96	17.86	17.02	16.32	15.70	15.14	14.93	13.67
14	21.89	20.13	19.00	18.13	17.40	16.77	16.20	15.98	14.67
15	23.10	21.30	20.13	19.24	18.49	17.84	17.25	17.02	15.67
16	24.31	22.46	21.26	20.34	19.58	18.90	18.29	18.06	16.67
17	25.50	23.61	22.39	21.44	20.66	19.97	19.34	19.10	17.67
18	26.70	24.76	23.51	22.54	21.74	21.03	20.38	20.14	18.67
19	27.88	25.91	24.63	23.64	22.81	22.09	21.43	21.18	19.67
20	29.07	27.05	25.74	24.73	23.89	23.15	22.47	22.22	20.67

Note: The basis for this table is the Poisson distribution. The 37 percent risk of incorrect acceptance column is provided for the convenience of those auditors that used previous MUS sampling formula guidance in developing policies and procedures.

Table C-4
Alternative MUS Sample Size Determination Using Expansion Factors

Risk of Incorrect Acceptance (%)	Factor
1	1.90
5	1.60
10	1.50
15	1.40
20	1.30
25	1.25
30	1.20
37	1.15
50	1.10

C.6 Previous versions of this guide used the preceding table to illustrate a formula approach for determining an MUS sample size for statistical sampling using expansion factors. This method is explained here using the example in chapter 6.

C.7 If the auditor expects misstatements, and the auditor is not using the table approach (table 4-5 or table C-1) or a formula approach using table C-2, but using a formula approach along with the expansion factors (table C-4, "Alternative MUS Sample Size Determination Using Expansion Factors"), he or she would reduce the tolerable misstatement by the expected misstatement, adjusted for the expansion factor appropriate for the desired assurance, and then proceed to determine sample size using the same approach described when zero misstatements are expected.

$$\text{Sample Size} = \frac{\text{Population Recorded Amount} \times \text{Confidence Factor}}{\text{Tolerable Misstatement} - (\text{Expected Misstatement} \times \text{Expansion Factor})}$$

C.8 As an example of the method using expansion factors, an auditor using MUS might have assessed tolerable misstatement as $15,000 and the desired risk of incorrect acceptance as 5 percent. In addition, the auditor may expect approximately $3,000 of misstatement in the population to be sampled. The expected effect of the misstatements is subtracted from the $15,000 tolerable misstatement. That effect is calculated by multiplying the expected misstatement, in this case $3,000, by an appropriate expansion factor. Table C-4 provides approximate expansion factors for some commonly used risks of incorrect acceptance. It gives an approximate expansion factor of 1.6 for a 5 percent risk of incorrect acceptance; therefore, the effect is $4,800 ($3,000 × 1.6). The auditor subtracts the $4,800 effect from the $15,000 tolerable misstatement and divides the resulting $10,200 ($15,000 − $4,800) by the appropriate confidence factor for applications in which no misstatements are expected, in this case a confidence factor of 3. The sampling interval in this example is $3,400

($10,200 ÷ 3). Therefore, for the population's recorded amount of $500,000, the sample size is computed to be 147 ($500,000 ÷ $3,400).

C.9 This sample size formula described is an approximation of the more accurate method used to compute the sample sizes in table 4-5 (table C-1). When zero misstatement is expected, this formula and the table give identical sample sizes. For low to moderate expected misstatement, the expansion factor formula gives sample sizes that are a bit smaller than the table. When expected misstatement is high—say, 40 percent or more of tolerable misstatement—the formula tends to result in sample sizes that exceed those in the table. In some cases, the excess is significant. The accuracy of the expansion factor formula approximation also varies with the risk of incorrect acceptance.

Appendix D

Ratio of Desired Allowance for Sampling Risk of Incorrect Rejection to Tolerable Misstatement

D.1 Table D-1, "Ratio of Desired Allowance for Sampling Risk of Incorrect Rejection to Tolerable Misstatement," is derived from *Statistical Auditing* by Donald Roberts (New York: AICPA, 1978) and is used in connection with the classical variables sampling guidance discussed in chapter 7, "Classical Variables Sampling." For further information on the theory underlying this measure of the risk of incorrect rejection, see pages 41–43 in *Statistical Auditing*.[1]

Table D-1

Ratio of Desired Allowance for Sampling Risk of Incorrect Rejection to Tolerable Misstatement

Risk of Incorrect Acceptance (One Sided)	Risk of Incorrect Rejection (One Sided)			
	0.10	0.05	0.025	0.005
0.010	0.355	0.414	0.457	0.525
0.025	0.395	0.456	0.500	0.567
0.050	0.437	0.500	0.543	0.610
0.075	0.470	0.533	0.576	0.641
0.100	0.500	0.562	0.604	0.667
0.150	0.552	0.613	0.654	0.713
0.200	0.603	0.661	0.699	0.753
0.250	0.655	0.709	0.743	0.792
0.300	0.709	0.758	0.788	0.830
0.350	0.768	0.810	0.835	0.869
0.400	0.834	0.866	0.885	0.910
0.450	0.910	0.929	0.939	0.953
0.500	1.000	1.000	1.000	1.000

Note: The basis for this table is the normal distribution.

---

[1] As described in *Statistical Auditing* by Donald Roberts (New York: AICPA, 1978), this table is based on the approach illustrated throughout this guide where the auditor accepts the population as not materially misstated unless there is evidence to the contrary (the positive approach). An equivalent, and sometimes a preferable approach (the negative approach), is where the auditor rejects the population as being materially misstated unless there is evidence to the contrary. The auditor using this latter approach would need to use a different table to relate the risks of incorrect acceptance and incorrect rejection than the one illustrated here.

Appendix E
Multilocation Sampling Considerations

E.1 This appendix deals with situations where the auditor has decided to select a sample of locations from a population of items at more than 1 location (for example, receivables at an entity's 200 locations). Further, the auditor intends to sample or perform other procedures at the locations selected for the sample. This appendix does not address the broader issues of planning, scoping, and executing multilocation audits.[1]

E.2 Auditors of multilocation entities may face additional sampling considerations beyond those encountered when applying audit sampling to a single population. The auditor may face such considerations when applying tests of controls or substantive tests of details. Common audit situations where such considerations may apply include inventories, fixed assets, or receivables that are in different locations.

E.3 In some cases it is feasible for the auditor to obtain sufficient evidence about all the locations by selecting one overall sample (for example, selecting from centralized records or visiting all locations). For example, the locations, although separate, might be in close proximity to each other, or audit resources may be readily available for all locations from which sample items might be selected. Generally, the audit strategy may be to first select any items or locations of greater risk for examination. Auditors also generally consider the nonsampling risks that may be introduced in some situations where the quality of evidence may differ when not visiting a location, such as examining original documentation and speaking directly to personnel.

E.4 In some cases, the auditor may be able to aggregate the populations of various locations and select an audit sample from the combined population, without further consideration of the location of the items selected for the sample. In this case, the sampling considerations are the same as applying sampling concepts to all locations. This approach generally produces the smallest overall sample size to meet the auditor's test objectives, but may require the auditor to perform procedures at many locations.

E.5 When it is not feasible to obtain the evidence centrally or visit all the locations, the auditor will generally select some locations from which to obtain audit evidence. In such cases, the auditor will generally first select those items or locations of greater risk or size for individual examination. If the auditor cannot select enough locations or items with this procedure to satisfy his or her audit objectives regarding the aggregate population, a sample of the remaining locations and a subsample of items from those locations may be selected to obtain the necessary assurance.

E.6 When a sample of locations is selected and a sample of items is selected from each location, the sampling risk from such a design consists of two risks: (*a*) risks associated with the examination of less than 100 percent of the locations (sometimes called *selection risk*), and (*b*) risks associated with

---

[1] Paragraph .43 of AU-C section 600, *Special Considerations—Audits of Group Financial Statements (Including the Work of Component Auditors)* (AICPA, *Professional Standards*), requires the auditor to assess the sufficiency and appropriateness of evidence obtained. A technique such as the one illustrated here may be helpful in making this assessment.

examining less than 100 percent of the items of interest at the locations visited (sometimes called the *condition detection risk*).

E.7 The total risk of the overall sampling plan is a combination of these two risks. For example, if the *selection risk* of the plan is 10 percent (for example, a 90 percent confidence level of identifying misstatements or deviations at a significantly misstated location), and there is also a 10 percent risk of not detecting a significant misstatement at a location selected, while the probabilities are not additive, there is approximately an overall 19 percent risk[2] that the plan will not be effective in detecting the pattern of misstatement exceeding tolerable misstatement.

E.8 In the determination of the overall extent of testing, fewer locations could be visited, increasing *selection risk* associated with locations. However, for a given overall confidence, this would ordinarily require more testing (accepting less risk of not detecting the error condition) to be performed at the locations visited. The auditor needs to visit enough locations and do enough work at each location to achieve the desired objective. Some auditors set minimum sample sizes for the number of locations to visit and number of items to test at each location.

E.9 When the auditor selects a sample of locations and then performs testing for each location, the auditor first evaluates the results of the sample for each location selected. If deviations or misstatements are found, the auditor considers whether those misstatements or deviations are likely in locations not visited. When evaluating sample results, the auditor considers whether the sample results might indicate a condition or pattern that might not support the assumptions used in developing the plan, indicating need for further evidence regarding the misstatements in the population. The auditor then aggregates the results of tests across all locations and assesses whether the desired assurance has been obtained from the procedures.

E.10 When statistical sampling is used, the auditor may need to consult with a sampling specialist to establish an appropriate sampling plan for the engagement circumstances. Statistical formulas can be used to project sample results from the sample results at the locations.[3]

---

[2] Formula: (90 percent Assurance × 90 percent Assurance = 81 percent Overall Confidence).

[3] Extensions of the concepts in the appendix are included in L. Graham, J. C. Bedard, and R. Cleary, Working Paper "The Multilocation Audit Issue" (Bentley University, 2012).

Appendix F

Mapping and Summarization of Changes — Clarified Auditing Standards

This appendix maps the extant[1] AU sections to the clarified AU-C sections. As a result of the Auditing Standards Board's (ASB's) Clarity Project, all extant AU sections have been modified. In some cases, individual AU sections have been revised into individual clarified standards. In other cases, some AU sections have been grouped together and revised as one or more clarified standards. In addition, the ASB revised the AU section number order established by Statement on Auditing Standards No. 1, *Responsibilities and Functions of the Independent Auditor* (AICPA, *Professional Standards*, AU sec. 110), to follow the same number order used in International Standards on Auditing (ISAs) for all clarified AU sections for which there are comparable ISAs. The clarified standards are effective for periods ending on or after December 15, 2012. Early adoption is not permitted.

Although the Clarity Project was not intended to create additional requirements, some revisions have resulted in changes that may require auditors to make adjustments in their practices. To assist auditors in the transition process, these changes have been organized into the following four types:

- Substantive changes
- Primarily clarifying changes
- Primarily formatting changes
- Standards not yet issued in the Clarity Project

This appendix identifies those AU-C sections associated with these four types of changes.

Substantive Changes

Substantive changes are considered likely to affect the firms' audit methodology and engagements because they contain *substantive* or *other changes*, defined as having one or both of the following characteristics:

- A change or changes to an audit methodology that may require effort to implement
- A number of small changes that, although not individually significant, may affect audit engagements

Primarily Clarifying Changes

Primarily clarifying changes are intended to explicitly state what may have been implicit in the extant standards, which, over time, resulted in diversity in practice.

(continued)

---

[1] The term *extant* is used throughout this appendix in reference to the standards that are superseded by the clarified standards.

Primarily Formatting Changes

Primarily formatting changes from the extant standards do not contain changes that expand the extant sections in any significant way and may not require adjustments to current practice.

Standards Not Yet Issued in the Clarity Project

Standards not yet issued in the Clarity Project contain the remaining sections that are in exposure or have not yet been reworked.

The preface of this guide and the Financial Reporting Center at www.aicpa.org/FRC/ provide more information about the Clarity Project. You can also visit www.aicpa.org/sasclarity.

Extant AU Sections Mapped to the Clarified AU-C Sections

Extant AU Section		AU Section Super-seded	New AU-C Section		Type of Change
110	Responsibilities and Functions of the Independent Auditor	All	200	Overall Objectives of the Independent Auditor and the Conduct of an Audit in Accordance With Generally Accepted Auditing Standards [1]	Primarily formatting changes
120	Defining Professional Requirements in Statements on Auditing Standards	All			
150	Generally Accepted Auditing Standards	All			
161	The Relationship of Generally Accepted Auditing Standards to Quality Control Standards	All	220	Quality Control for an Engagement Conducted in Accordance With Generally Accepted Auditing Standards	Primarily clarifying changes
201	Nature of the General Standards	All	200	Overall Objectives of the Independent Auditor and the Conduct of an Audit in Accordance With Generally Accepted Auditing Standards [1]	Primarily formatting changes
210	Training and Proficiency of the Independent Auditor	All			
220	Independence	All			
230	Due Professional Care in the Performance of Work	All			
311	Planning and Supervision	All except paragraphs .08–.10	300	Planning an Audit	Primarily formatting changes
		Paragraphs .08–.10	210	Terms of Engagement	Primarily clarifying changes

(continued)

Extant AU Sections Mapped to the Clarified AU-C Sections—*continued*

Extant AU Section		AU Section Superseded	New AU-C Section		Type of Change
312	Audit Risk and Materiality in Conducting an Audit	All	320	Materiality in Planning and Performing an Audit	Primarily formatting changes
			450	Evaluation of Misstatements Identified During the Audit	Primarily formatting changes
314	Understanding the Entity and Its Environment and Assessing the Risks of Material Misstatement	All	315	Understanding the Entity and Its Environment and Assessing the Risks of Material Misstatement	Primarily formatting changes
315	Communications Between Predecessor and Successor Auditors	All except paragraphs .03–.10 and .14	510	Opening Balances— Initial Audit Engagements, Including Reaudit Engagements	Primarily clarifying changes
		Paragraphs .03–.10 and .14	210	Terms of Engagement	Primarily clarifying changes
316	Consideration of Fraud in a Financial Statement Audit	All	240	Consideration of Fraud in a Financial Statement Audit	Primarily formatting changes
317	Illegal Acts by Clients	All	250	Consideration of Laws and Regulations in an Audit of Financial Statements	Substantive changes
318	Performing Audit Procedures in Response to Assessed Risks and Evaluating the Audit Evidence Obtained	All	330	Performing Audit Procedures in Response to Assessed Risks and Evaluating the Audit Evidence Obtained	Primarily formatting changes

Mapping and Summarization of Changes

Extant AU Sections Mapped to the Clarified AU-C Sections—*continued*

Extant AU Section		AU Section Superseded	New AU-C Section		Type of Change
322	The Auditor's Consideration of the Internal Audit Function in an Audit of Financial Statements	All	Planned to be issued as AU-C section 610	The Auditor's Consideration of the Internal Audit Function in an Audit of Financial Statements	Standards not yet issued in the Clarity Project
324	Service Organizations	All	402	Audit Considerations Relating to an Entity Using a Service Organization	Primarily clarifying changes
325	Communicating Internal Control Related Matters Identified in an Audit	All	265	Communicating Internal Control Related Matters Identified in an Audit	Substantive changes
326	Audit Evidence	All	500	Audit Evidence	Primarily formatting changes
328	Auditing Fair Value Measurements and Disclosures	All	540	Auditing Accounting Estimates, Including Fair Value Accounting Estimates, and Related Disclosures [2]	Primarily formatting changes
329	Analytical Procedures	All	520	Analytical Procedures	Primarily formatting changes
330	The Confirmation Process	All	505	External Confirmations	Primarily clarifying changes
331	Inventories	All	501	Audit Evidence— Specific Considerations for Selected Items [3]	Primarily clarifying changes
332	Auditing Derivative Instruments, Hedging Activities, and Investments in Securities	All	501	Audit Evidence— Specific Considerations for Selected Items [3]	Primarily clarifying changes

(continued)

Extant AU Sections Mapped to the Clarified AU-C Sections—*continued*

Extant AU Section		AU Section Super-seded	New AU-C Section		Type of Change
333	Management Representations	All	580	Written Representations	Primarily formatting changes
334	Related Parties	All	550	Related Parties	Substantive changes
336	Using the Work of a Specialist	All	620	Using the Work of an Auditor's Specialist	Primarily Clarifying Changes
337	Inquiry of a Client's Lawyer Concerning Litigation, Claims, and Assessments	All	501	Audit Evidence—Specific Considerations for Selected Items [3]	Primarily clarifying changes
339	Audit Documentation	All	230	Audit Documentation	Primarily formatting changes
341	The Auditor's Consideration of an Entity's Ability to Continue as a Going Concern	All	Planned to be issued as AU-C section 570	Going Concern (in exposure)	Standards not yet issued in the Clarity Project
342	Auditing Accounting Estimates	All	540	Auditing Accounting Estimates, Including Fair Value Accounting Estimates, and Related Disclosures [2]	Primarily formatting changes
350	Audit Sampling	All	530	Audit Sampling	Primarily formatting changes
380	The Auditor's Communication With Those Charged With Governance	All	260	The Auditor's Communication With Those Charged With Governance	Primarily formatting changes
390	Consideration of Omitted Procedures After the Report Date	All	585	Consideration of Omitted Procedures After the Report Release Date	Primarily formatting changes

AAG-SAM APP F

Extant AU Sections Mapped to the Clarified AU-C Sections—*continued*

Extant AU Section		AU Section Superseded	New AU-C Section		Type of Change
410	Adherence to Generally Accepted Accounting Principles	All	700	Forming an Opinion and Reporting on Financial Statements [4]	Substantive changes
420	Consistency of Application of Generally Accepted Accounting Principles	All	708	Consistency of Financial Statements	Primarily clarifying changes
431	Adequacy of Disclosure in Financial Statements	All	705	Modifications to the Opinion in the Independent Auditor's Report [5]	Primarily formatting changes
504	Association With Financial Statements	All	N/A	Withdrawn	

(continued)

Extant AU Sections Mapped to the Clarified AU-C Sections—*continued*

Extant AU Section		AU Section Superseded	New AU-C Section		Type of Change
508	Reports on Audited Financial Statements	Paragraphs .01–.11, .14–.15, .19–.32, .35–.52, .58–.70, and .74–.76	700	Forming an Opinion and Reporting on Financial Statements [4]	Substantive changes
			705	Modifications to the Opinion in the Independent Auditor's Report [5]	Primarily formatting changes
			706	Emphasis-of-Matter Paragraphs and Other-Matter Paragraphs in the Independent Auditor's Report [6]	Substantive changes
		Paragraphs .12–.13	600	Special Considerations—Audits of Group Financial Statements (Including the Work of Component Auditors)	Substantive changes
		Paragraphs .16–.18 and .53–.57	708	Consistency of Financial Statements	Primarily clarifying changes
		Paragraphs .33–.34	805	Special Considerations—Audits of Single Financial Statements and Specific Elements, Accounts, or Items of a Financial Statement	Primarily clarifying changes
		Paragraphs .71–.73	560	Subsequent Events and Subsequently Discovered Facts [7]	Primarily formatting changes

Mapping and Summarization of Changes

Extant AU Sections Mapped to the Clarified AU-C Sections—*continued*

Extant AU Section		AU Section Superseded	New AU-C Section		Type of Change
530	Dating of the Independent Auditor's Report	Paragraphs .01–.02	700	Forming an Opinion and Reporting on Financial Statements [4]	Substantive changes
		Paragraphs .03–.08	560	Subsequent Events and Subsequently Discovered Facts [7]	Primarily formatting changes
532	Restricting the Use of an Auditor's Report	All	905	Alert That Restricts the Use of the Auditor's Written Communication	Primarily clarifying changes
534	Reporting on Financial Statements Prepared for Use in Other Countries	All	910	Financial Statements Prepared in Accordance With a Financial Reporting Framework Generally Accepted in Another Country	Primarily clarifying changes
543	Part of Audit Performed by Other Independent Auditors	All	600	Special Considerations— Audits of Group Financial Statements (Including the Work of Component Auditors)	Substantive changes
544	Lack of Conformity With Generally Accepted Accounting Principles	All	800	Special Considerations— Audits of Financial Statements Prepared in Accordance With Special Purpose Frameworks [8]	Primarily clarifying changes
550	Other Information in Documents Containing Audited Financial Statements	All	720	Other Information in Documents Containing Audited Financial Statements	Primarily formatting changes

(continued)

Extant AU Sections Mapped to the Clarified AU-C Sections—*continued*

Extant AU Section		AU Section Super-seded	New AU-C Section		Type of Change
551	Supplementary Information in Relation to the Financial Statements as a Whole	All	725	Supplementary Information in Relation to the Financial Statements as a Whole	Primarily formatting changes
552	Reporting on Condensed Financial Statements and Selected Financial Data	All	810	Engagements to Report on Summary Financial Statements	Primarily clarifying changes
558	Required Supplementary Information	All	730	Required Supplementary Information	Primarily formatting changes
560	Subsequent Events	All	560	Subsequent Events and Subsequently Discovered Facts [7]	Primarily formatting changes
561	Subsequent Discovery of Facts Existing at the Date of the Auditor's Report	All			

Extant AU Sections Mapped to the Clarified AU-C Sections—*continued*

Extant AU Section		AU Section Superseded	New AU-C Section		Type of Change
623	Special Reports	Paragraphs .19–.21	806	Reporting on Compliance With Aspects of Contractual Agreements or Regulatory Requirements in Connection With Audited Financial Statements	Primarily formatting changes
		Paragraphs .01–.10 and .22–.34	800	Special Considerations— Audits of Financial Statements Prepared in Accordance With Special Purpose Frameworks [8]	Primarily clarifying changes
		Paragraphs .11–.18	805	Special Considerations— Audits of Single Financial Statements and Specific Elements, Accounts, or Items of a Financial Statement	Primarily clarifying changes
625	Reports on the Application of Accounting Principles	All	915	Reports on Application of Requirements of an Applicable Financial Reporting Framework	Primarily formatting changes
634	Letters for Underwriters and Certain Other Requesting Parties	All	920	Letters for Underwriters and Certain Other Requesting Parties	Primarily formatting changes
711	Filings Under Federal Securities Statutes	All	925	Filings With the U.S. Securities and Exchange Commission Under the Securities Act of 1933	Primarily formatting changes

(continued)

Extant AU Sections Mapped to the Clarified AU-C Sections—*continued*

Extant AU Section		AU Section Superseded	New AU-C Section		Type of Change
722	Interim Financial Information	All	930	Interim Financial Information	Primarily formatting changes
801	Compliance Audits	All	935	Compliance Audits	Primarily formatting changes
901	Public Warehouses—Controls and Auditing Procedures for Goods Held	All	501	Audit Evidence—Specific Considerations for Selected Items [3]	Primarily clarifying changes

Legend:

[n] Bracketed number indicates a clarity standard that supersedes more than one extant AU section.

The AICPA has developed an Audit Risk Alert to assist auditors and members in practice prepare for the transition to the clarified standards. It has been organized to give you the background information on the development of the clarified standards and to identify the new requirements and changes from the extant standards. Check out the Audit Risk Alert *Understanding the Clarified Auditing Standards* (product no. ARACLA12P), which is available in the AICPA store on www.cpa2biz.com.

Appendix G

Glossary

G.1 This glossary summarizes definitions of the terms related to audit sampling used in this guide. It does not contain definitions of common audit terms. Related terms are shown in parentheses.

allowance for sampling risk (precision). A measure of the difference between a sample estimate (projection) and the tolerable rate of deviation or tolerable misstatement at a specified sampling risk.

alpha risk. See **risk of incorrect rejection**.

attribute. Any characteristic that is either present or absent in a sampling unit. In tests of controls, the presence or absence of evidence of the application of a specified control is sometimes referred to as an attribute.

attributes sampling. Statistical sampling that reaches a conclusion about a population in terms of a rate of occurrence.

audit risk. The risk that the auditor expresses an inappropriate audit opinion when the financial statements are materially misstated. Audit risk is a function of the *risks of material misstatement* and detection risk.

audit sampling (sampling). The selection and evaluation of less than 100 percent of the population of audit relevance such that the auditor expects the items selected (the sample) to be representative of the population and, thus, likely to provide a reasonable basis for conclusions about the population. In this context, *representative* means that evaluation of the sample will result in conclusions that, subject to the limitations of sampling risk, are similar to those that would be drawn if the same procedures were applied to the entire population.

basic precision. In monetary unit sampling, the minimum allowance for sampling risk. It equals the allowance for sampling risk when no misstatements are found in the sample.

beta risk. See **risk of incorrect acceptance**.

biased selection. A selection that is not selected in such a way to be expected to be representative of the population from which it was selected (for example, selecting only smaller value invoices for examination). See **representative**.

binomial distribution. In probability theory and statistics, the binomial distribution is the discrete probability distribution of the number of successes in a sequence of n independent draws, each of which yields success with probability p. Because the probability p is unchanged by each draw, it is an accurate description of sampling *with replacement* before the next draw. In large populations, the binomial distribution can yield an approximation of the hypergeometric distribution when the sample size is less than 10 percent of the population size.

block sample. This is a sample consisting of contiguous sampling units. Many blocks are generally needed to form a sample that can be expected to be representative.

cell sampling. A form of monetary unit sampling or probability proportional to size sample selection where the population is divided into sampling intervals and a sample selection is made from each sampling interval (cell). Some monetary unit sampling evaluation techniques also are based on cell theory.

classical variables sampling. A statistical sampling approach that measures sampling risk using the variation of the underlying characteristic of interest. This approach includes methods such as mean-per-unit, ratio estimation, difference estimation, and a classical form of probability proportional to size estimation.

cluster sample. See **block sample**.

confidence level (reliability). The complement of the risk of incorrect acceptance. The measure of probability associated with a sample interval.

control risk. The risk that a misstatement that could occur in an assertion about a class of transaction, account balance, or disclosure and that could be material, either individually or when aggregated with other misstatements, will not be prevented, or detected and corrected, on a timely basis by the entity's internal control.

cumulative monetary amount (CMA) sampling. See **monetary unit sampling**.

decision model. A rule used to make a conclusion about a population based on a sample taken from it.

detection risk. The risk that the procedures performed by the auditor to reduce audit risk to an acceptably low level will not detect a misstatement that exists and that could be material, either individually or when aggregated with other misstatements.

difference estimation. A classical variables sampling technique that uses the average difference between individual audited amounts and individual recorded amounts to estimate the total audited amount (or the total misstatement) of a population and an allowance for sampling risk.

dollar-unit sampling (DUS). See **monetary unit sampling**.

expansion factor. A factor used in the calculation of sample size in a monetary unit sampling application if misstatements are expected.

expected population deviation rate. An anticipation of the deviation rate in the entire population. It is used in determining an appropriate sample size for an attributes sample.

factual misstatements. Misstatements about which there is no doubt. These were called *known* misstatements in prior auditing literature.

field. See **population**.

haphazard sample. A sample consisting of sampling units selected without any conscious bias (that is, without any special reason for including or omitting items from the sample). It does not consist of sampling units selected in a careless manner and is selected in a manner that can be expected to be representative of the population.

hypergeometric distribution. In probability theory and statistics, the hypergeometric distribution is a discrete probability distribution that describes the probability associated with a number of occurrences of a particular outcome in a sequence of n draws from a finite population (for example, without replacement of the selected item before the next item is drawn).

hypothesis testing. A decision model to test the reasonableness of an amount by assessing whether sample data is consistent or otherwise with statements made about the population.

inherent risk. The susceptibility of an assertion about a class of transaction, account balance, or disclosure to a misstatement that could be material, either individually or when aggregated with other misstatements, before consideration of any related controls.

judgmental misstatements. Differences arising from the judgments of management concerning accounting estimates that the auditor considers unreasonable or the selection or application of accounting policies that the auditor considers inappropriate.

known misstatements. See **factual misstatements.**

likely misstatement (most likely misstatement). In audit sampling, likely misstatement is the direct projection, or best estimate of the sample result when extrapolated to the population from which the sample was drawn. See also **projected misstatement**.

logical unit. The balance or transaction that includes the selected dollar in a monetary unit sample.

mean-per-unit approach. A classical variables sampling technique that projects the sample average to the total population by multiplying the sample average by the total number of items in the population.

misstatement. A difference between the amount, classification, presentation, or disclosure of a reported financial statement item and the amount, classification, presentation, or disclosure that is required for the item to be presented fairly in accordance with the applicable financial reporting framework. Misstatements can arise from fraud or error. (Also see definitions of factual misstatement, judgmental misstatement and projected misstatement.)

monetary unit sampling (MUS). A form of variables sampling based on attributes sampling theory that uses probability proportional to size sample selection. Sometimes called *dollar unit sampling*.

nonsampling risk. The risk that the auditor reaches an erroneous conclusion for any reason not related to sampling risk.

nonstatistical sampling. A sampling technique for which the auditor considers sampling risk in evaluating an audit sample without using statistical theory to measure that risk.

normal distribution. The normal distribution is a continuous probability distribution, applicable in many fields. It may be defined by two parameters: the mean (*average*, μ) and variance (*variability*, σ^2), respectively. The standard normal distribution is the normal distribution with a mean of zero and a variance of one. Carl Friedrich Gauss became associated with this set of distributions when he analyzed astronomical data using them, and defined the equation of its probability density function. It is often called the bell curve because the graph of its probability density resembles a bell. It is often used in the application of classical variables sampling techniques. The normal distribution can yield an approximation of the binomial distribution when the occurrence probability is close to 50 percent.

performance materiality. The amount or amounts set by the auditor at less than materiality for the financial statements as a whole to reduce to an appropriately low level the probability that the aggregate of uncorrected and undetected misstatements exceeds materiality for the financial statements as a whole. If applicable, *performance materiality* also refers to the amount or amounts set by the auditor at less than the materiality level or levels for particular classes of transactions, account balances, or disclosures. Performance materiality is to be distinguished from tolerable misstatement.

point estimate. Most likely amount of the population characteristic based on the extrapolation of the sample results. Also known as the *likely misstatement* or *best estimate amount*.

Poisson distribution. In probability theory and statistics, the Poisson distribution is a discrete probability distribution that expresses the probability of a number of events occurring in a fixed period of time if these events occur with a known average rate, and are independent of the time since the last event. As applied in auditing, it yields a reasonable approximation of the hypergeometric distribution when the population occurrence rate and the sampling fraction (sample size ÷ population) are both less than 10 percent, conditions common in many auditing populations.

population. The entire set of data from which a sample is selected and about which the auditor wishes to draw conclusions.

population for sampling purposes. The population for sampling purposes excludes individually significant items that the auditor has decided to examine 100 percent or other items that will be tested separately.

precision. See **allowance for sampling risk**.

probability proportional to size (PPS) sampling. A sample selection procedure that selects items for the sample in proportion to their relative size, usually their monetary amounts. Monetary unit sampling uses this method to select the sample. There is also a probability proportional to size sampling estimation procedure that is based on classical variables sampling techniques. This latter technique requires enough misstatements in the sample in order to form appropriate statistical confidence limits. Both monetary unit sampling and probability proportional to size estimation samples are selected on a proportional to size basis.

projected misstatements. The auditor's best estimate of misstatements in populations, involving the projection of misstatements identified in audit samples to the entire population from which the samples were drawn. Also see **likely misstatement**.

random sample. A sample selected so that every combination of the same number of items in the population has an equal probability of selection.

Glossary

ratio estimation. A classical variables sampling technique that uses the ratio of audited amounts to recorded amounts in the sample to estimate the total dollar amount of the population and an allowance for sampling risk.

reciprocal population. See **related population**.

related population. A population containing items that may be missing from or understated in the population of interest. For example, in testing for completeness of accounts payable (the population of interest), the auditor may identify a related population of subsequent payments and select from that population; if that related population is overstated, the population of interest is understated.

reliability level. See **confidence level**.

representative. Evaluation of the sample will result in conclusions that, subject to the limitations of sampling risk, are similar to those that would be drawn if the same procedures were applied to the entire population.

In many contexts in sampling, representative conveys the sense that the sample results are believed to correspond, at the stated risk level, to what would have been obtained had the auditor examined all items in the population in the same way as examined in the sample. *Correspond* does not mean that the projected misstatement from the sample will exactly equal the misstatement in the population (which the auditor does not know). Rather a sample is considered representative if it is free from selection bias. Statistical samples are designed to be representative, with the stated confidence that the true population misstatement is measured by the confidence interval. Nonstatistical samples are generally selected in a way that the auditor expects them to be representative. Representative relates to the total sample, not to individual items in the sample. Also, representative does not relate to the sample size, but to how the sample was selected. The sample is generally expected to be representative only with respect to the occurrence rate or incidence of misstatements, not their specific nature. A sample misstatement due to an unusual circumstance may nevertheless be indicative of other unusual misstatements in the population.

risk of incorrect acceptance (beta risk or type II misstatement). The risk that the sample supports the conclusion that the recorded account balance is not materially misstated when the account balance is materially misstated.

risk of incorrect rejection (alpha risk or type I misstatement). The risk that the sample supports the conclusion that the recorded account balance is materially misstated when the account balance is not materially misstated.

risk of material misstatement (RMM). The risk that the financial statements are materially misstated prior to the audit. This consists of two components, described as follows at the assertion level:

> **inherent risk.** The susceptibility of an assertion about a class of transaction, account balance, or disclosure to a misstatement that could be material, either individually or when aggregated with other misstatements, before consideration of any related controls.

control risk. The risk that a misstatement that could occur in an assertion about a class of transaction, account balance, or disclosure and that could be material, either individually or when aggregated with other misstatements, will not be prevented, or detected and corrected, on a timely basis by the entity's internal control.

risk of overreliance. The risk that the auditor will conclude the controls are more effective than they are. Also referred to as beta risk, type II, risk of incorrect acceptance, risk of assessing control risk too low. See **sampling risk**.

risk of underreliance. The risk that the auditor will conclude the controls are less effective than they actually are. Also referred to as alpha risk, type I, risk of incorrect rejection, risk of assessing control risk too high. See **sampling risk.**

sample. Items selected from a population to reach a conclusion about the population as a whole.

sampling distribution. The set of all possible outcomes of a sample from a population. Some sampling distributions are exact, such as the hypergeometric distribution, which compute the probability of a specific (attribute based) outcome from a population of any known size, given a random sample and known population characteristics. The binomial distribution is often an effective approximation to the hypergeometric distribution and may be used when the population is large. The Poisson distribution is another attribute based approximation method that may be used when the estimated misstatement or deviation rate and the proportion of the population being sampled is small. Classical variables sampling often relies on theoretical distributions such as the normal distribution or Student T distribution to compute the statistical confidence limits and can consider standard deviation. These latter distributions are based on large-sample theory.

sampling error. See **allowance for sampling risk**.

sampling risk. The risk that the auditor's conclusion based on a sample may be different from the conclusion if the entire population were subjected to the same audit procedure. Sampling risk can lead to two types of erroneous conclusions:

    a. In the case of a test of controls, that controls are more effective than they actually are, or in the case of a test of details, that a material misstatement does not exist when, in fact, it does. The auditor is primarily concerned with this type of erroneous conclusion because it affects audit effectiveness and is more likely to lead to an inappropriate audit opinion.

    b. In the case of a test of controls, that controls are less effective than they actually are, or in the case of a test of details, that a material misstatement exists when, in fact, it does not. This type of erroneous conclusion affects audit efficiency because it would usually lead to additional work to establish that initial conclusions were incorrect.

sampling unit. The individual items constituting a population.

sequential sampling. A sampling plan for which the sample is selected in several steps, with each step conditional on the results of the previous steps. The development of a valid plan that considers the risks of allowing for multiple stages of sampling generally requires specialized tables or specialist assistance, and cannot be directly inferred from single stage sampling plans or tables.

standard deviation. A measure of the dispersion among the respective amounts of a particular characteristic as measured for all items in the population for which a sample estimate is developed.

standard error. The standard deviation of the sampling distribution of a statistic.

statistic. A numerical characteristic of a sample. For example, the sample mean and variance.

statistical sampling. An approach to sampling that has the following characteristics:

  a. Random selection of the sample items
  b. The use of an appropriate statistical technique to evaluate sample results, including measurement of sampling risk

  A sampling approach that does not have characteristics *a* and *b* is considered nonstatistical sampling.

stop-or-go sampling. See **sequential sampling**.

stratification. The process of dividing a population into subpopulations, each of which is a group of sampling units that have similar characteristics. Stratification may be used to focus procedures on risk areas or to reduce variability in sampling populations.

substantive procedure. An audit procedure designed to detect material misstatements at the assertion level. Substantive procedures comprise.

  a. tests of details (classes of transactions, account balances, and disclosures) and
  b. substantive analytical procedures.

systematic random sampling. A method of selecting a sample in which every *n*th item is selected using one or more random starts. When the first item is selected using judgment from the interval, the method is termed systematic sampling

tainting. In a monetary-unit sample, the percentage of misstatement present in a logical unit. It is usually expressed as the ratio of the amount of misstatement in the item to the item's recorded amount.

test of controls. An audit procedure designed to evaluate the operating effectiveness of controls in preventing, or detecting and correcting, material misstatements at the assertion level.

tolerable misstatement. A monetary amount set by the auditor in respect of which the auditor seeks to obtain an appropriate level of assurance that the monetary amount set by the auditor is not exceeded by the actual misstatement in the population.

tolerable rate of deviation. A rate of deviation set by the auditor in respect of which the auditor seeks to obtain an appropriate level of assurance that the rate of deviation set by the auditor is not exceeded by the actual rate of deviation in the population.

type I error. See **risk of incorrect rejection**

type II error. See **risk of incorrect acceptance**

uncorrected misstatements. Misstatements that the auditor has accumulated during the audit and that have not been corrected.

universe. See **population**.

variables sampling. A sampling method that reaches a conclusion on the monetary amounts of a population. It includes monetary unit sampling and classical variables sampling techniques.

Appendix H

Schedule of Changes Made to the Text From the Previous Edition

As of March 1, 2012

This schedule of changes identifies areas in the text and footnotes of this guide that have been changed from the previous edition. Entries in the table of this appendix reflect current numbering, lettering (including those in appendix names), and character designations that resulted from the renumbering or re-ordering that occurred in the updating of this guide.

Reference	Change
General	Guidance related to the clarified auditing standards (Statement on Auditing Standards Nos. 122–125) has been incorporated throughout this guide. See appendix F, "Mapping and Summarization of Changes–Clarified Auditing Standards," for a mapping of the extant standards to the clarified AU-C sections.
Notice to readers	Deleted for the passage of time.
Preface	Updated.
Paragraphs I.21–.22	Added for clarification.
Appendix F	Added.
Former appendix G	Deleted for the passage of time.
Former appendix H	Deleted for the passage of time.
Index	Added.

Index

1

100 PERCENT EXAMINATION
- Audit sampling 1.10, 2.34
- Classical variables sampling 7.44
- Monetary unit sampling ...... 6.16, 6.35–.41, ... 6.58
- Nonstatistical sampling case study .... 5.12–.13 (table 5-1), 5.16
- Projected misstatement to population 4.77–.86
- Substantive tests of details 4.12, 4.39

A

ACCEPTABLE RISK LEVELS
- Nonstatistical sampling case study 5.08
- Substantive tests of details 4.33–.38

ACCOUNT BALANCES
- Accounts-receivable balances .... 4.14, ... 5.02–.16
- Audit sampling 1.11, 2.30–.32
- Gross margins, sample planning and 4.58
- Monetary unit sampling 6.06, 6.55
- Nonstatistical sampling case study 5.01–.16
- Sampling vs. nonsampling audit procedures 1.21–.22
- Substantive tests of details .... 4.02, 4.07, ... 4.58

ACCOUNTS-RECEIVABLE BALANCES
- Monetary unit sampling 6.06, 6.08
- Nonstatistical sampling case study 5.02–.16
- Substantive tests of details 4.14

ACHIEVED PRECISION, CLASSICAL VARIABLES SAMPLING 7.25

ACTUAL RATE OF DEVIATION, TESTS OF CONTROLS 3.54–.57

ADJUSTED GROSS EXPOSURE, MAGNITUDE OF DEFICIENCY ESTIMATIONS 3.89

ADJUSTMENT, PROPOSED 4.90, 7.27

AICPA PROFESSIONAL STANDARDS
- Clarified auditing standards, mapping and summarization of changes .... App. F

ALLOWANCE FOR SAMPLING RISK. See also precision
- Classical variables sampling ...... 7.06, 7.23, ... 7.28, 7.31–.34
- Incorrect rejection .... App. D

ALPHA RISK. See incorrect rejection, risk of

ANALYTICAL PROCEDURES
- Auditing procedures 1.08
- Classical variables sampling ...... 7.05, 7.34, ... 7.45
- Monetary unit sampling 6.49–.50
- Nonstatistical sampling case study 5.07–.08, 5.10
- Substantive tests of details .... 4.40–.42 .... (table 4-2), 4.56

"AS OF" DATE, TESTS OF CONTROLS 3.13

ASB (AUDITING STANDARDS BOARD), CLARITY PROJECT .... App. F

ASSERTIONS
- Nonstatistical sampling case study 5.07, 5.10
- Sample design for 4.59

ASSET VERIFICATION, SAMPLING VS. NONSAMPLING PROCEDURES 1.24

ASSURANCES
- Audit reliability and confidence level 1.27
- Nonstatistical sampling case study 5.08, 5.10, 5.12
- Sample design for assertions 4.59
- Substantive tests of details 4.72–.74
- Tests of controls 3.44–.45

ATTEST ENGAGEMENTS, TESTS OF CONTROLS 3.10

ATTRIBUTES SAMPLING
- Applications 2.41
- Monetary unit sampling 6.02, 6.09, 6.30
- Procedure 2.35–.37
- Statistical sampling tables .... App. A
- Substantive tests of details 4.16
- Tests of controls 3.32, 3.65

AUDIT RISK MODEL
- Classical variables sampling 7.45
- Substantive tests of details ... 4.39–.47, 4.68

AUDIT SAMPLING CHARACTERISTICS 1.01–.29
- Best practices 1.03
- Characteristics not evaluated 1.16–.17
- Control tests, exclusion of 1.13–.14
- Cutoff tests excluded from 1.12
- Defined 1.04, 4.06
- Excluded procedures 1.06, 1.09
- Extrapolation not intended 1.15, 4.09
- IT systems, automated controls, testing of 1.19
- Nonsampling procedures vs. 1.22–.25
- Population 1.11
- Precision 1.29
- Procedures excluded from 1.06–.19
- Reliability or confidence level 1.27

AAG-SAM AUD

AUDIT SAMPLING CHARACTERISTICS—continued
- Representative sampling 1.05
- Sampling risk 1.28
- Terminology 1.26
- Untested balances, exclusion of 1.18

AUDIT SAMPLING PROCESS 2.01–.55
- Attributes sampling 2.35–.37
- Audit tests 2.08–.14
- Balance sheet and income statement 2.42–.43
- Continuing professional education ...2.45–.48
- Differences from sampling in other professions 2.03–.06
- Documentation 4.105–.108
- Efficiency and appropriateness of 2.34, 4.43, 5.10
- Evaluation of 2.07
- Guidelines 2.49
- Implementation of procedures 2.44–.55, 3.25–.28
- Non-audit sampling compared with 2.03–.06
- Nonstatistical sampling 2.22–.29, 3.01–.98, 5.01–.16
- Nonstatistical sampling case study 5.01–.16
- Objectives of 2.33
- Planning process of 2.30–.34
- Population sequence estimation errors ...3.69
- Process 2.01–.55
- Purpose and nature of 2.02
- Risk 2.15–.21
- Sample selection methods 3.29–.36
- Sample size 2.28
- Specialists, use of 2.50–.51
- Statistical sampling 2.22–.29, 3.01–.98
- Substantive tests of details 4.01–.108
- Supervision and review 2.52–.55
- Tests of controls 3.01–.98
- Variables sampling 2.38–.41

AUDIT STRATEGY
- Nonstatistical sampling case study 5.10
- Tests of controls 3.12

AUDIT TESTS. See also tests of controls
- Account balances and income states 2.42
- Attributes sampling 2.36–.37
- Audit sample supervision and review 2.54–.55
- Audit sampling 2.08–.14
- Classical variables sampling 7.05, 7.34
- Dual-purpose tests 2.12–.14
- Monetary unit sampling 6.06
- Nonstatistical sampling case study 5.05, 5.10
- Sampling risk at level of 4.93–.100
- Substantive tests of details 4.01–.108
- Substantive tests of details compared with 4.56
- Variables sampling 2.38–.40

AUDITING PROCEDURES
- Analytical 1.08
- Audit sampling process 2.01–.55
- Characteristics not evaluated 1.16–.17
- Evidence sources 1.02
- Exclusion from audit sampling 1.06–.19
- Inquiry and observation 1.07
- Items included in 1.10, 4.11–.12
- Nonstatistical sampling case study 5.04
- Sampling procedure documentation 3.96–.98
- Sampling vs. nonsampling procedures 1.20–.25
- Substantive tests of details 4.04
- Tests of controls 3.16–.17, 3.71–.72
- Untested balances 1.18

AUDITING STANDARDS
- Standards clarification, mapping and summarization of changes ....App. F

AUTOMATED CONTROLS
- IT systems, tests of 1.19
- Tests of controls 1.14
- Walkthrough procedures 3.27

B

BALANCE SHEETS
- Audit sampling procedures 2.42–.43
- Materiality measures 4.57
- Projected misstatement to population 4.81
- Substantive tests of details 4.02, 4.07

BASIC PRECISION, MONETARY UNIT SAMPLING 6.39, 6.44, 6.47

"BEST ESTIMATE," MONETARY UNIT SAMPLING 6.63

BEST PRACTICES
- Audit sampling procedures and planning 2.31
- Haphazard sampling 3.34
- Overreliance risk, multiple controls 3.47
- Substantive tests of details 4.06
- Tests of controls 3.21

BETA RISK. See incorrect acceptance, risk of

BIASED SELECTION
- Audit sampling characteristics 1.05
- Classical variables sampling 7.31, 7.35–.39
- Substantive tests of details 4.09–.10, 4.103

BLOCK SAMPLING, TESTS OF CONTROLS 3.35–.36

C

CASH RECEIPTS, NONSTATISTICAL SAMPLING CASE STUDY 5.03

CELL SAMPLING
- Monetary unit sampling 6.16
- Tests of controls 3.32

Index

CLARIFIED AUDITING STANDARDS ..... **1.01,** .... **App. F**

CLARITY PROJECT, AUDITING STANDARDS BOARD (ASB) .... **App. F**

CLASS OF TRANSACTIONS
- Audit sampling 1.11, 2.30–.32
- Monetary unit sampling 6.08
- Projected misstatement to population 4.81
- Sampling vs. nonsampling audit procedures 1.21

CLASSICAL VARIABLES SAMPLING . 7.01–.48
- Advantages 7.04
- Allowance for sampling risk .... App. D
- Applications 2.41
- Basic techniques 7.02
- Case study 7.40–.48
- Design considerations 7.04, 7.12
- Disadvantages 7.05
- Examples 2.40
- Information requirements 7.14–.15
- Method classification 7.07–.10
- Method selection 7.11–.15
- Monetary unit sampling 2.38–.41
- Monetary unit sampling vs. .... 6.03–.08, ... 7.04–.06, 7.24, 7.27
- Sample results evaluation 7.25–.39, 7.48
- Sample size determination ... 7.05, 7.16–.24, ... 7.34–.35, App. D
- Statistical approach 7.03–.06
- Substantive tests of details .... 4.16, 4.30, ... 4.62, 4.84

COMBINED ATTRIBUTES/VARIABLES SAMPLING, MONETARY UNIT SAMPLING **6.02**

COMMUNICATION PROCEDURES, TESTS OF CONTROLS .... **3.17, See also documentation requirements**

CONFIDENCE FACTORS
- Audit sampling 1.27
- Classical variables sampling 7.15
- Monetary unit sampling .... 6.23–.31 .... (table 6-1), 6.36–.41 (table 6-2), 6.56
- Monetary unit sampling tables .... App. C
- Substantive tests of details 4.72–.74

CONFIDENCE LEVEL
- Audit sampling 1.27
- Classical variables sampling 7.20, 7.22
- Monetary unit sampling 6.21–.22, 6.63
- Substantive tests of details ... 4.55, 4.72–.74

CONFIRMATION PROCESS
- Monetary unit sampling 6.07, 6.59
- Nonstatistical sampling case study 5.04, 5.07, 5.10, 5.12

CONTAINMENT **4.101–.104**

CONTINUING PROFESSIONAL EDUCATION (CPE), AUDIT SAMPLING PROCEDURES **2.45–.48**

CONTROL RISK
- Classical variables sampling 7.45
- Monetary unit sampling 6.54
- Nonstatistical sampling case study 5.05

CONTROLLED SOURCES, TESTS OF CONTROLS **3.20**

CONTROLS
- Audit sampling procedures 2.09–.10
- Automated controls, IT systems tests 1.19
- Changes in 3.08
- Dual-purpose tests 2.12–.14
- Magnitude of deficiency in 3.85–.90
- Performance of, audit sampling 1.14, 3.08
- Tests of controls. See tests of controls
- Tolerable rate of deviation determination 3.51

COST-BENEFIT ANALYSIS **2.23, 7.18**

CPE. See continuing professional education (CPE)

CREDIT BALANCES **4.07, 5.02, 6.08**

CUMULATIVE MONETARY AMOUNTS (CMA), MONETARY UNIT SAMPLING **6.02, 6.07**

CUSTOM CONTRACTUAL TERMS, BALANCE SHEET AND INCOME STATEMENT SAMPLING **2.43**

CUSTOMER BALANCES **5.12, 6.58**

CUT-OFF TESTS
- Attributes sampling 2.37
- Audit sampling 1.12
- Classical variables sampling 7.43
- Nonstatistical sampling case study 5.07

D

DEBIT BALANCES **4.07, 5.02, 5.07**

DECISION MODELS, STATISTICAL SAMPLING **2.23**

DETECTION RISK, SUBSTANTIVE TESTS OF DETAILS **4.35–.36**

DEVIATION CONDITIONS. See also tolerable rate of deviation
- Attributes statistical sampling tables ... App. A
- Defined 3.06
- Examples 3.91–.92
- Magnitude of deficiency 3.85–.90
- Sample extension and 3.82–.84
- Sampling documentation 3.97

DEVIATION RATE CALCULATION
- Sampling documentation 3.97
- Tests of controls ..... 3.74, 3.80–.84, App. B

DIFFERENCE ESTIMATION
- Classical variables sampling .... 7.05, 7.07, ... 7.09, 7.13–.15, 7.43
- Monetary unit sampling 6.07
- Substantive tests of details .... 4.62, 4.84, ... 4.86

DIRECT PROJECTION
- Classical variables sampling 7.30
- Deviation rate calculation 3.74
- Monetary unit sampling 6.49, 6.63

DISTRIBUTION OF MISSTATEMENTS, SUBSTANTIVE TESTS OF DETAILS **4.86**

DOCUMENTATION REQUIREMENTS
- Audit sample supervision and review .... 2.53, ... 3.96–.98
- Projected misstatement to population 4.78
- Sampling procedure 4.105–.108
- Tests of controls .... 3.04, 3.24, 3.67, ... 3.96–.98
- Unused/inapplicable documents 3.68
- Voided documents 3.67

DOLLAR-UNIT SAMPLING **6.02, 6.16**

DOLLAR-VALUE ESTIMATION
- Nonstatistical sampling case study .... 5.13 (table 5-1)
- Projected misstatements 4.83
- Substantive tests of details ... 4.04, 4.19–.22 ....(table 4-1)
- Variables sampling 2.38–.39

DOUBLE CREDIT, MAGNITUDE OF DEFICIENCY ESTIMATIONS **3.86**

DUAL-PURPOSE TESTS
- Audit sampling 2.12–.14
- Monetary unit sampling 6.09

E

EFFECTIVENESS OF CONTROLS, TESTS OF CONTROLS **3.13, 3.27**

ESTIMATION ERRORS
- Population sequences, tests of control ... 3.69
- Substantive tests of details 4.52, 4.84

EVIDENCE SOURCES
- Audit sampling 2.06
- Auditing procedures 1.02
- Classical variables sampling 7.35–.39
- Monetary unit sampling 6.49–.51
- Small populations and infrequently operating controls 3.62, 3.65
- Substantive tests of details .... 4.03, 4.36, ... 4.79, 4.90
- Tests of controls 3.04–.05, 3.25, 3.62
- Tolerable rate of deviation determination 3.49
- Unused/inapplicable documents 3.68

EXPANSION FACTORS, MONETARY UNIT SAMPLING **6.27**

EXPECTED MISSTATEMENTS
- Classical variables sampling .. 7.19–.20, 7.27
- Monetary unit sampling ...... 6.07, 6.25–.31, ... 6.56
- Monetary unit sampling tables .... App. C

EXPECTED MISSTATEMENTS—continued
- Nonstatistical sampling case study 5.05, 5.08–.09
- Projected misstatement to population 4.79
- Substantial tests of details .... 4.60–.61, ... 4.72, 4.74

EXPECTED POPULATION DEVIATION RATE, TESTS OF CONTROLS **3.55–.58**

EXTENDED SAMPLE SIZE
- Classical variables sampling .... 7.34–.35, ... 7.37
- Deviation conditions 3.82–.84
- Monetary unit sampling 6.49–.51, 6.55

EXTERNAL CONFIRMATIONS, SUBSTANTIVE TESTS OF DETAILS **4.14, 4.91**

F

FACTUAL MISSTATEMENT
- Classical variables sampling 7.47–.48
- Monetary unit sampling .... 6.35–.41, 6.43 ....(table 6-3), 6.48, 6.63
- Nonstatistical sampling case study 5.05, 5.13–.16
- Substantive tests of details 4.77–.86
- Test level sampling risk 4.93–.100

FIFO. See first in, first out (FIFO) inventory

FINANCIAL STATEMENTS
- Clarified auditing standards ...... 1.01, App. F
- Classical variables sampling 7.38
- Nonstatistical sampling case study ...... 5.05, ... 5.10, 5.15–.16

FINITE POPULATION CORRECTION FACTOR, CLASSICAL VARIABLES SAMPLING **7.05**

FIRST IN, FIRST OUT (FIFO) INVENTORY **6.08, 7.06**

FIXED SAMPLE SIZE ...... **3.62, 3.83, App. B**

FIXED-ASSET ADDITIONS **1.22–.23, 6.06**

FIXED-INTERVAL SELECTION, MONETARY UNIT SAMPLING **6.13–.14**

FORMULA APPROACH, NONSTATISTICAL SAMPLING
- Monetary unit sampling 6.23–.31
- Substantial tests of details 4.72–.74

FRAUD RISK, NONSTATISTICAL SAMPLING CASE STUDY **5.04**

G

GENERALLY ACCEPTED ACCOUNTING PRINCIPLES (GAAP), BALANCE SHEET AND INCOME STATEMENT SAMPLING **2.43**

GLOSSARY OF TERMS .... **App. G**

GROSS MARGINS, SAMPLE PLANNING AND 4.58
GROSS VALUATION OF ACCOUNTS, NONSTATISTICAL SAMPLING CASE STUDY 5.07

H

HAPHAZARD SAMPLING METHOD
· Nonstatistical sampling case study 5.11
· Substantive tests of details 4.17
· Tests of controls 3.29, 3.33–.34
HYPERGEOMETRIC DISTRIBUTION, POPULATION SIZE ESTIMATION 3.59
HYPOTHESIS TESTING
· Substantive tests of details 4.04
· Variables sampling 2.38

I

IMPLICIT SAMPLING INTERVAL, SUBSTANTIVE TESTS OF DETAILS 4.85
INAPPLICABLE/UNUSED DOCUMENTS 3.68, 4.76
INCOME STATEMENTS
· Audit sampling procedures 2.42–43
· Materiality measures 4.57
INCORRECT ACCEPTANCE, RISK OF
· Acceptable risk levels 4.33–.38
· Classical variables sampling 7.08,
.... 7.15–.17, 7.20–.24, 7.28–.29,
.... 7.34, 7.44–.48, App. D
· Monetary unit sampling 6.20, 6.23–.24,
.... 6.39, 6.49
· Monetary unit sampling tables .... App. C
· Nonstatistical sampling case study 5.08–.09
· Performance materiality and tolerable misstatement 4.50
· Substantive tests of details ...4.26, 4.33–.38,
.... 4.40 (table 4-2), 4.68–.69 (table 4-5)
INCORRECT REJECTION, RISK OF
· Acceptable risk levels 4.34–.38
· Allowance for sampling risk .... App. D
· Classical variables sampling 7.20–.24,
.... 7.28–.29, 7.44–.48, App. D
· Substantive tests of details 4.43–.47
INCREMENTAL ALLOWANCE, MONETARY UNIT SAMPLING, UPPER LIMIT ON MISSTATEMENTS WITH TAINTINGS 6.44–.48, 6.62
INFORMATION REQUIREMENTS, CLASSICAL VARIABLES SAMPLING 7.14–.15
INFORMATION TECHNOLOGY (IT) SYSTEMS
· Automated controls, tests of 1.19, 3.27
· Tests of controls 3.13

INFREQUENTLY OPERATING CONTROLS, TESTS OF CONTROL 3.62–.63
.............................(table 3-5)
INHERENT RISK, NONSTATISTICAL SAMPLING CASE STUDY 5.05
INQUIRY, AUDITING PROCEDURES 1.07
INTEREST CHARACTERISTICS, SUBSTANTIVE TESTS OF DETAILS 4.05
INTERIM SAMPLE RESULTS
· Substantive tests of details 4.92
· Tests of controls during 3.11
INTERNAL CONTROLS
· Classical variables sampling 7.41
· Nonstatistical sampling case study 5.03
· Tests of controls 3.05, 3.12–.13,
.................................3.17, 3.78
INTERPOLATION TECHNIQUES, MONETARY UNIT SAMPLING 6.30
INTERVAL ESTIMATION
· Monetary unit sampling 6.18–.19, 6.23,
.... 6.38, 6.43, 6.49, 6.56–.57
· Tolerable rate of deviation 3.53
INVENTORY TEST COUNTS
· Audit sampling procedures 2.27
· Auditor's test counts 1.15
· Classical variables sampling 7.06, 7.08,
.... 7.15, 7.40–.48
· Monetary unit sampling 6.06, 6.08
INVESTMENT TESTS, MONETARY UNIT SAMPLING 6.06
IT. See information technology (IT) systems
ITEM SELECTION
· Assertions, sample design for 4.59
· Audited and recorded amounts, difference 4.84
· Auditing procedures 1.10, 4.11–.12
· Classical variables sampling 7.04, 7.35,
.... 7.44
· Monetary unit sampling 6.12, 6.17, 6.50
· Nonstatistical sampling case study 5.06,
................................5.10, 5.13, 5.16
· Substantive tests of details ...4.11–.12, 4.18,
............................4.27, 4.71, 4.76
· Tests of controls 3.21, 3.72
INTERIM SAMPLE RESULTS
· Substantive tests of details 4.92

J

JUDGMENTAL DIFFERENCE, MONETARY UNIT SAMPLING 6.63

L

LAST IN, FIRST OUT (LIFO) INVENTORY 4.04, 6.08, 7.06

AAG-SAM LAS

LAWS OF PROBABILITY, STATISTICAL
SAMPLING 2.23
LIFO. See last in, first out (LIFO) inventory
LIKELIHOOD CRITERION, TESTS OF
CONTROLS 3.85, 3.87
LOANS RECEIVABLE CONFIRMATION,
MONETARY UNIT SAMPLING 6.06
LOGICAL UNITS, MONETARY UNIT SAMPLING
- Defined 6.09
- Project misstatements with taintings 6.43
- Sample selection 6.11, 6.13,
.... 6.15–.16, 6.18
- Upper limit on misstatements 6.41–.42
- Upper limit on misstatements
  with taintings 6.45–.48

M

MAGNITUDE OF CONTROL DEFICIENCY,
TESTS OF CONTROLS 3.85–.90
MAGNITUDE OF MISSTATEMENTS,
SUBSTANTIVE TESTS OF
DETAILS 4.86
MANUAL CONTROL ACTIVITIES
- Attributes sampling 2.36
- Monetary unit sampling 6.18–.19
MATERIALITY
- Classical variables sampling 7.20, 7.31,
.... 7.38
- Monetary unit sampling 6.51, 6.54, 6.57
- Nonstatistical sampling case
  study 5.05, 5.16
- Performance materiality, substantive
  tests of details 4.50–.59
MEAN PER UNIT ESTIMATION
- Classical variables sampling 7.07–.08,
.... 7.12–.15, 7.43
- Substantive tests of details 4.62
MINIMUM SAMPLE SIZE, CLASSICAL
VARIABLES SAMPLING 7.05, 7.24
MISSTATEMENT. See also factual
misstatement; projected misstatement
- Adjustment proposals, evidence 4.90
- Aggregation and assessment 4.99–.102
- Classical variables sampling 7.30–.31,
.... 7.35–.39, 7.43–.48
- Factual 4.77–.86, 4.93–.100, 5.05,
.... 5.13–.16, 6.35–.41, 6.43 (table 6-3),
.... 6.48, 6.63, 7.47–.48
- Monetary unit sampling 6.07, 6.25–.31,
.... 6.35–.41, 6.50–.61
- Nonstatistical sampling case
  study .... 5.13–.16 (table 5-1)
- Not projected 4.101–.104
- Projected misstatement 4.77–.86, 4.90,
.... 4.93–.102, 5.11, 5.13–.14
- Projection to population 4.77–.86
- Qualitative analysis 4.87–.89
- Unprojected misstatements 4.101–.104

MISSTATEMENTS NOT
PROJECTED 4.101–.104
MONETARY UNIT SAMPLING
(MUS) 6.01–.63
- Advantages of 6.05–.06
- Applications 6.03
- Attributes/variables sampling 6.02
- Case study .... 6.53–.62 (table 6-5)
- Classical variables sampling 2.38–.41
- Classical variables sampling vs. 6.03–.08,
.... 7.04–.06, 7.24, 7.27
- Disadvantages of 6.07–.08
- Formula method, expected
  misstatements 6.26–.31
- Formula method, unexpected
  misstatements .... 6.23–.25 (table 6-1)
- Incorrect acceptance, five percent
  risk .... 6.36–.41 (table 6-2)
- Judgmental difference 6.63
- Less than 100 percent misstatements ... 6.42
- Nonstatistical sampling case study 5.13
- 100 percent misstatements 6.35–.41
- Projected misstatement evaluation 6.35,
.... 6.60–.62 (table 6-5)
- Proportional to size sampling 6.01
- Qualitative analysis 6.52
- Quantitative analysis 6.49–.51
- Results evaluation 6.32–.52,
.... 6.59–.62 (table 6-5)
- Sample design 6.55
- Sample selection criteria 6.11–.19,
.... 6.56–.58
- Sample size 4.68, 6.20–.22
- Sample tables .... App. C
- Sampling unit defined 6.09–.10
- Statistical method selection 6.04
- Substantive tests of details 4.16, 4.27,
.... 4.29, 4.62, 4.68, 4.71,
.... 4.82, 4.85–.86
- Taintings, projected misstatement
  evaluation .... 6.43 (table 6-3)
- Tests of controls 3.32
- Upper limit on misstatements, when taintings
  occur .... 6.44–.48 (table 6-4)

MULTILOCATION SAMPLING
CONSIDERATIONS .... App. E
MULTIPLE INDEPENDENT CONTROLS,
OVERRELIANCE RSK AND 3.47

N

NEGATIVE BALANCES
- Classical variables sampling 7.04
- Monetary unit sampling 6.07, 6.17
NEGATIVE CONFIRMATIONS
- Substantive tests of details 4.91
NONSAMPLING RISK, AUDIT SAMPLING
PROCESS 2.20–.21

AAG-SAM LAW

Index

NONSTATISTICAL AUDIT SAMPLING
- Account balance or class of transactions 2.32
- Attributes statistical sampling tables ... App. A
- Case study 5.01–.16
- Classical variables sampling 7.27
- Design 2.24
- Deviations in 3.91–.92
- Haphazard sampling 3.29, 3.33–.34
- Inventory count 2.27
- Procedures for 2.22–.29
- Sample results evaluation 5.12–.16
- .... (table 5-1)
- Sample risk evaluation 3.78
- Sample selection methods 3.29–.36
- Sample size 2.28, 5.08–.11
- Sampling risk 2.28–.29
- Substantive tests of details 4.01–.108
- Tests of controls 3.01–.98

NORMAL DISTRIBUTION THEORY
- Classical variables sampling ...... 7.02, 7.05, ... 7.08–.10
- Substantive tests of details 4.16

O

OBSERVATION, AUDITING PROCEDURES 1.07

ONE-SIDED UPPER BOUNDS
- Classical variables sampling ..... 7.21, App. D
- Monetary unit sampling 6.07

OUTSIDE CONSULTANTS, AUDIT SAMPLING PROCEDURES 2.51

OVERRELIANCE, RISK OF
- Attributes statistical sampling tables ... App. A
- Deviation and sample extension 3.84
- Expected population deviation rate 3.58
- Monetary unit sampling ...... 6.07, 6.33–.34, ... 6.36–.41
- Multiple controls 3.47
- Population size effect on sample size 3.60
- .... (table 3-4)
- Sample size .... 3.41 (table 3-1)
- Sampling procedure documentation 3.96
- Tests of controls 3.38–.47, 3.53, 3.60
- Tolerable rate of deviation and 3.46
- Tolerable rate of deviation and 3.52
- .... (table 3-2)

OVERSTATED BALANCES
- Classical variables sampling 7.06, 7.42
- Monetary unit sampling ...... 6.03, 6.07–.08, ... 6.25, 6.32–.37, 6.48, 6.53, ... 6.55, 6.59, 6.63
- Nonstatistical sampling case study .... 5.12 (table 5-1)

P

PERCENTAGE OF MISSTATEMENT, MONETARY UNIT SAMPLING 6.43
- .... (table 6-3), 6.56

PERFORMANCE MATERIALITY
- Classical variables sampling 7.41
- Monetary unit sampling ...... 6.51, 6.53–.54, ... 6.56–.57
- Nonstatistical sampling case study 5.05
- Substantive tests of details ... 4.11, 4.50–.59
- .... (table 4-3)

PERFORMANCE OF CONTROLS
- Sampling documentation 3.96–.98
- Tests of controls 3.06

PHYSICAL REPRESENTATION OF POPULATIONS
- Substantive tests of details .... 4.09–.10, ... 4.103
- Tests of controls 3.18–.21

PILOT SAMPLING, SUBSTANTIVE TESTS OF DETAILS 4.30

PLANNING PROCESS
- Audit sampling 2.30–.34
- Nonstatistical sampling case study 5.03
- Sample plan performance, tests of control 3.66–.72
- Tests of controls 3.12

POINT ESTIMATE
- Classical variables sampling 7.27, 7.34
- Deviation rate calculation 3.74
- Monetary unit sampling 6.49, 6.63
- Nonstatistical sampling case study 5.13
- Substantive tests of details 4.85

POPULATION CHARACTERISTICS
- Audit sampling 1.11, 2.04–.06
- Classical variables sampling .... 7.03–.06, ... 7.15
- Estimation errors 3.69
- Expected population deviation rate 3.55–.58
- Misstatement projections .... 4.77–.86, ... 5.06, 5.13
- Monetary unit sampling .... 6.05–.08, ... 6.14, 6.49–.61
- Non-auditing populations 2.04–.06
- Substantive tests of details .... 4.06–.07, ... 4.27–.32, 4.57–.59
- Tests of controls 3.07–.10, 3.14–.21
- Tolerable misstatements and 4.57–.59

POPULATION COMPLETENESS
- Substantive tests of details 4.08–.10
- Tests of controls 3.18–.21

POPULATION SIZE
- Classical variables sampling 7.17–.24
- Sample size, limited effect 3.60–.61
- .... (table 3-4)
- Small populations, tests of control 3.62–.63
- Substantive tests of details 4.62–.74
- .... (tables 4-4 and 4-5)
- Tests of controls 3.15, 3.59–.61

AAG-SAM POP

PRECISION. See also allowance for sampling risk
- Audit sampling 1.29
- Classical variables sampling .... 7.15, ... 7.20–.21, 7.25, 7.31–.34, ... 7.37, App. D
- Monetary unit sampling .... 6.39–.40, ... 6.48, 6.61
- Sample risk evaluation 3.78–.79
- Substantive tests of details ... 4.28, 4.55–.56

PRICING TESTS
- Classical variables sampling ...... 7.08, 7.40, ... 7.43–.48
- Monetary unit sampling 6.06, 6.08

PROBABILITY PROPORTIONAL TO SIZE (PPS) SELECTION
- Classical variables sampling 7.04
- Monetary unit sampling 6.01, 6.07
- Nonstatistical sampling case study 5.11, 5.13–.14
- Substantive tests of details ... 4.17–.18, 4.21, ... 4.71, 4.85

PROBABILITY WEIGHTING, SUBSTANTIVE TESTS OF DETAILS 4.21–.22

PROJECTED MISSTATEMENTS
- Classical variables sampling .... 7.15, ... 7.27–.31, 7.47–.48
- Evidence for adjustment 4.90
- Misstatements not projected 4.101–.104
- Monetary unit sampling .... 6.32–.41, ... 6.43–.49 (tables 6-3 and 6-4), ... 6.60 (table 6-5), 6.63
- Nonstatistical sampling case study 5.06, 5.13–.16
- Substantive tests of details ... 4.77–.86, 4.90
- Test level sampling risk 4.93–.100

Q

QUALITATIVE ANALYSIS
- Classical variables sampling 7.39
- Deviation rates 3.80–.81
- Magnitude of deficiency estimations 3.90
- Monetary unit sampling 6.52
- Substantive tests of details 4.87–.89

QUANTITATIVE ANALYSIS
- Classical variables sampling 7.43
- Monetary unit sampling 6.49–.51

R

RANDOM SAMPLING METHODS
- Classical variables sampling 7.08, 7.34
- Population sequence estimation errors ... 3.69
- Tests of controls 3.29–.32, 3.66, 3.69

RANDOM START, MONETARY UNIT SAMPLING 6.13, 6.18, 6.49

RANKED ESTIMATIONS, MONETARY UNIT SAMPLING RESULTS .... 6.46–.48 .... (table 6-4)

RATIO METHODS
- Classical variables sampling ...... 7.05, 7.07, ... 7.10, 7.13–.15, 7.21, 7.43, App. D
- Desired sampling risk .... App. D
- Monetary unit sampling 6.07, 6.56
- Nonstatistical sampling case study 5.06
- Substantive tests of details 4.86

REASONABLE BASIS STANDARD, CLASSICAL VARIABLES SAMPLING 7.27

RECIPROCAL POPULATIONS, MONETARY UNIT SAMPLING 6.07

RECLASSIFICATIONS, TOLERABLE MISSTATEMENT FOR 4.57

RECORDED AMOUNTS
- Classical variables sampling ...... 7.05, 7.09, ... 7.17, 7.30–.35, 7.39, 7.43–.44
- Disadvantages of, monetary unit sampling 6.07–.08
- Monetary unit sampling case study 6.57, 6.60
- PPS samples 6.01
- Projected misstatements 6.35
- Quantitative analysis 6.50–.61
- Sample results 6.32–.33
- Sample selection criteria ..... 6.13, 6.15–.16, ... 6.18–.19
- Sample size 6.10, 6.24, 6.30
- Upper limit on misstatements ...... 6.40, 6.43 .... (table 6-3), 6.48

REGRESSION ESTIMATION, CLASSICAL VARIABLES SAMPLING 7.07

RELATED POPULATION SELECTION, CLASSICAL VARIABLES SAMPLING 7.05

RELIABILITY FACTORS. See also confidence level
- Audit sampling 1.27
- Monetary unit sampling .... 6.23 (table 6-1)
- Substantive tests of details 4.72–.74

REPEATED PROCEDURES, TESTS OF CONTROLS 3.28

REPERFORMANCE, TESTS OF CONTROLS 1.13–14, 3.25

REPRESENTATIVE SAMPLING
- Audit sampling characteristics 1.05
- Classical variables sampling .... 7.31, ... 7.35–.39
- Substantive tests of details .... 4.09–.10, ... 4.103

REVENUE RECOGNITION
- Attributes sampling 2.37
- Nonstatistical sampling case study 5.03–.04

AAG-SAM PRE

Index

REVIEW, AUDIT SAMPLING PROCESS 2.52–.55

RISK
- Audit reliability and confidence level 1.27
- Audit sampling 2.15–.21
- Of incorrect acceptance ..... 4.26, 4.33–.38, ... 4.40 (table 4-2), 4.68 (table 4-5), 5.08
- Of incorrect rejection 4.34–.38, 4.43–.47
- Nonsampling risk 2.20–.21
- Nonstatistical sampling case study 5.04
- Sampling risk. See also sampling risk

RISK OF MATERIAL MISSTATEMENT
- Audit sampling 1.11, 2.17
- Classical variables sampling 7.45
- Incorrect acceptance 4.34–.38
- Monetary unit sampling 6.53–.56, 6.63
- Nonstatistical sampling case study 5.04–.05, 5.08, 5.14
- Sampling vs. nonsampling audit procedures 1.21, 1.24
- Substantive tests of details .... 4.02–.03, ... 4.05, 4.26, 4.33–.47 (table 4-2), ... 4.75, 4.100
- Tests of controls 3.11, 3.50, 3.95
- Tolerable rate of deviation determination 3.50
- Untested balances 1.18

S

SAMPLE AVERAGING, PERFORMANCE MATERIALITY 4.53

SAMPLE PLANNING PERFORMANCE
- Gross margins in 4.58
- Results evaluation 3.73–.95, 5.12–.16
- Sampling procedure documentation 3.96–.98
- Sequential sampling tables .... App. B
- Substantive tests of details .... 4.04, 4.58, ... 4.63–.76 (tables 4-4 and 4-5)
- Tests of controls 3.66–.95

SAMPLE RESULTS EVALUATION
- Classical variables sampling .... 7.25–.39, ... 7.48
- Monetary unit sampling ... 6.32–.52, 6.59–.62
- Sample planning 5.12–.16
- Substantive tests of details ... 4.77–.86, 4.92
- Tests of controls 3.11, 3.73–.95

SAMPLE SELECTION METHODS
- Monetary unit sampling ... 6.11–.19, 6.56–.58
- Substantive tests of details 4.17–.22
- Tests of controls 3.29–.36

SAMPLE SIZE DETERMINATION
- Classical variables sampling .... 7.05, ... 7.15–.24, 7.34–.35, 7.44–.48, App. D
- Deviations and extension of 3.82–.84
- Expected population deviation rate ...... 3.56 .... (table 3-3)
- Guidelines 3.65

SAMPLE SIZE DETERMINATION—continued
- Monetary unit sampling .... 4.68, 6.05, ... 6.20–.31
- Monetary unit sampling tables .... App. C
- Nonstatistical sampling case study 5.08–.11
- Overreliance and .... 3.41 (table 3-1)
- Population size, limited effect .... 3.60–.61 ....(table 3-4)
- Sequential or fixed sample size .... 3.62, App. B
- Substantive tests of details ... 4.23–.26, 4.28, ... 4.55, 4.62–.74 (tables 4-4 and 4-5)
- Tests of controls 3.37–.65

SAMPLING ERROR. See allowance for sampling risk

SAMPLING RISK
- Acceptable risk levels 4.33–.38
- Actual rate of deviation 3.54–.57
- Audit sampling process 2.19
- Classical variables sampling ...... 7.08, 7.10, ... 7.13, 7.23, 7.27–.34, ... 7.41–.48, App. D
- Defined 3.38
- Evaluation of 3.75–.79
- Monetary unit sampling ...... 6.32–.34, 6.39, ... 6.45–.48
- Multiple independent controls 3.47
- Nonstatistical vs. statistical sampling 2.28–.29
- Overreliance risk 3.38–.47, 3.53, 3.60
- Projected misstatement to population 4.77–.86
- Substantive tests of details .... 4.33–.38, ... 4.43–.47, 4.93–.100
- Test interruption due to 3.71
- Test level risk 4.93–.100
- Tests of controls ... 3.38–.45, 3.71, 3.75–.79
- Underreliance 1.28

SAMPLING UNITS
- Classical variables sampling ...... 7.09, 7.24, ... 7.34
- Inability to examine 3.72
- Monetary unit sampling .... 6.09–.10, ... 6.49–.51
- Population size and 3.62
- Substantive tests of details .... 4.08, ... 4.13–.14
- Tests of controls 3.18, 3.22–.24, 3.72

SEQUENTIAL SAMPLING
- Deviation and extension of 3.83–.84
- Tables .... App. B
- Tests of control .... 3.62, App. B

SIMPLE RANDOM SAMPLING, TESTS OF CONTROLS 3.29–.30, 3.66

SMALL POPULATIONS
- Classical variables sampling 7.05
- Monetary unit sampling 6.07
- Tests of control .... 3.62–.63 (table 3-5)

SOFTWARE REQUIREMENTS
- Classical variables sampling ...... 7.02, 7.22, ... 7.33, 7.42, 7.46–.47
- Statistical audit sampling 2.25–.26
- Substantive tests of details 4.30

SPECIALISTS, USE OF IN AUDIT SAMPLING 2.50–.51

STANDARD DEVIATION
- Classical variables sampling ...... 7.05, 7.33, ... 7.48
- Substantive tests of details 4.27

STATISTICAL AUDIT SAMPLING
- Account balance or class of transactions 2.32
- Attributes sampling tables .... App. A
- Classical variables sampling 7.03–.06
- Continuing professional education ... 2.45–.48
- Inventory count 2.27
- Monetary unit sampling .... 6.04–.08, ... 6.36–.41
- Monetary unit sampling tables ...... App. C
- Multilocation sampling issues .... App. E
- Outside consultants 2.51
- Procedures for 2.22–.29
- Sample selection methods 3.29–.36
- Sampling risk 2.28
- Sequential sampling tables .... App. B
- Substantive tests of details 4.01–.108
- Tests of controls 3.01–.98
- Training and expertise requirements 2.25

STATISTICAL THEORY, AUDIT SAMPLE SIZE 4.67

STOP-AND-GO SAMPLING, TESTS OF CONTROL 3.64

STRATIFIED SAMPLING
- Classical variables sampling .... 7.12–.19, ... 7.44, 7.48
- Monetary unit sampling 6.05
- Nonstatistical sampling case study 5.08
- Substantive tests of details .... 4.19–.22 .... (table 4-1), 4.29–.30, 4.71

STRATUM SUBDIVISIONS IN POPULATION
- Classical variables sampling .... 7.12, ... 7.14–.19, 7.24, 7.46
- Nonstatistical sampling case study ...... 5.06, ... 5.08, 5.11
- Substantive tests of details 4.28, 4.30

STRINGER BOUND UPPER LIMIT APPROACH, MONETARY UNIT SAMPLING RESULTS .... 6.46 (table 6-4)

STUDENT T DISTRIBUTION, CLASSICAL VARIABLES SAMPLING 7.05

SUBSTANTIVE PROCEDURES
- Audit sample supervision and review 2.55
- Audit sampling 2.11
- Classical variables sampling .... 7.30–.31, ... 7.34–.35, 7.45
- Expected amount of misstatement 4.61

SUBSTANTIVE PROCEDURES—continued
- Monetary unit sampling 6.04–.08, 6.49
- Nonstatistical sampling case study 5.07–.08
- Over and underreliance 3.39
- Substantive tests of details .... 4.04, 4.36, ... 4.40–.42, 4.56

SUBSTANTIVE TESTS OF DETAILS 4.01–.108
- Acceptable risk levels 4.33–.38
- Audit risk model 4.39–.47
- Audit sampling technique, selection criteria 4.15–.16
- Classical variables sampling 7.11–.15
- Confidence reliability factors .... 4.72 ....(table 4-6)
- Documentation of sample procedure 4.105–.108
- Evidence sufficiency, proposed adjustments 4.90
- Expected amount of misstatement 4.60–.61
- Generally 4.01
- Incorrect acceptance ..... 4.34–.42 (table 4-2)
- Incorrect rejection 4.43–.47
- Interim sample results 4.92
- Misstatement to population, projection 4.77–.87
- Negative confirmations 4.91
- Nonstatistical sampling case study 5.08
- Performance and evaluation using 4.03
- Performance materiality .... 4.50–.59 ....(table 4-3)
- Population characteristics 4.06–.07
- Population completeness 4.08–.10
- Population size effects 4.62
- Population variability 4.27–.32
- Purpose 4.02
- Qualitative factors 4.87–.89
- Risk of material misstatements ...... 4.40–.47 ....(table 4-2)
- Sample plan performance 4.75–.76
- Sample results evaluation 4.77–.86
- Sample selection criteria 4.17–.22
- Sample size determination .... 4.23–.26, ... 4.63–.74
- Sampling risk, test levels 4.93–.100
- Sampling unit defined 4.13–.14
- Significant item identification 4.11–.12
- Stratification, examples .... 4.19–.22 ....(table 4-1)
- Test objectives 4.04–.05
- Tolerable misstatement .... 4.48–.59 ....(table 4-3)
- Unprojected misstatements 4.101–.104

SUPERVISION, AUDIT SAMPLING PROCESS 2.52–.55

SYSTEMATIC SAMPLING METHODS
- Monetary unit sampling 6.05, 6.13, 6.57
- Tests of controls 3.31–.32

Index

T

TAINTING, MONETARY UNIT SAMPLING
- Disadvantages 6.07
- Project misstatements 6.35, 6.43–.48
- Results evaluation 6.32–.48
- Sample size 6.21
- Upper limit on projected
  misstatements 6.44–.48

TEST DECK, AUDIT SAMPLING 1.14

TEST LEVEL SAMPLING RISK, SUBSTANTIVE TESTS OF DETAILS 4.93–.100

TEST OF DETAILS (TD)
- Risk of .... 4.40–.42 (table 4-2)
- Substantive. See substantive tests of details

TESTS OF CONTROLS .... 3.01–.98, See also audit tests
- Actual rate of deviation 3.54–.57
- Adjusted gross exposure 3.89
- Applications 3.04
- Attest engagement 3.10
- Attributes sampling 2.36
- Attributes statistical sampling tables ... App. A
- Audit sample supervision and review 2.55
- Audit sampling exclusions 1.13–.14
- Audit sampling procedures 2.09–.10
- Block sampling 3.35–.36
- Classical variables sampling 7.05, 7.41
- Control changes 3.08
- Control deficiency, potential magnitude
  assessment 3.85–.90
- Control deviations, sample
  extension 3.81–.84
- Controlled source, selection from 3.20
- Controls assurance 3.79
- Controls effectiveness
  assessment...................... 3.38–.45
- Design guidelines 3.03
- Deviation conditions 3.06
- Deviation rate calculation 3.74, 3.80–.84
- Dual-purpose tests 2.12–14
- Effectiveness of controls, assessment ... 3.13
- Evidence requirements 3.11, 3.46
- Examples 3.91–.92
- Expected population deviation
  rate 3.55–.57
- Extrapolation not intended 1.15
- Fixed sample size approach 3.64
- Haphazard sampling 3.33–.34
- Inability to examine items 3.72
- Infrequently operating controls 3.62–.63
- Initial testing 3.14
- Interim periods 3.11
- Internal controls 3.05, 3.12
- Interruption of test 3.71
- IT systems, automated controls 1.19
- Magnitude of exposure, upper limit
  measurement 3.94
- Missing items 3.21
- Monetary unit sampling 6.09, 6.30

TESTS OF CONTROLS—continued
- Nonstatistical audit sampling 3.01–.98,
  .... 5.05, 5.08, 5.10
- Operating effectiveness
  assessment 3.25, 3.28
- Overall conclusions 3.95
- Overreliance risk, assessment of ... 3.38–.47,
  .... 3.53, 3.60
- Period covered by test 3.11–.13
- Physical representation, unit
  selection from 3.18–.21
- Population characteristics 3.07–.10,
  .... 3.15–.17
- Population completeness 3.18–.21
- Population sequence estimation
  errors 3.69
- Population size 3.15, 3.59–.61
- Precision 3.54, 3.77–.78
- Procedures 3.93
- Qualitative analysis of deviations 3.80–.81
- Reliance assessment 3.27
- Sample results evaluation 3.73–.95
- Sample selection method 3.29
- Sample size determination 3.37–.65
- Sampling plan performance 3.66–.72
- Sampling procedure
  documentation 3.96–.98
- Sampling risk 1.28, 3.75–.79
- Sampling unit characteristics 3.22–.24
- Sampling unit selection 3.18
- Sequential sample size approach 3.64
- Sequential sampling tables .... App. B
- Significance of deficiency 3.90
- Simple random sampling 3.29–.30
- Small populations 3.62–.63
- Special situations 3.70
- Statistical sampling case study 3.01–.98
- Stop-or-go sampling 3.64
- Substantive procedures assessment 3.26
- System changes and 3.09
- Systematic sampling 3.31–.32
- Test objectives 3.02–.05
- Tolerable rate of deviation 3.43, 3.45–.46,
  .... 3.48–.56, 3.76
- Underreliance, risk of 3.38–.41
- Unused or inapplicable documents 3.68
- Upper-limit approach
  - deviation rate 3.35–.54, 3.77, 3.88
  - exposure magnitude 3.94
- Usage underestimation 3.16
- Voided documents 3.67
- Walkthroughs 3.25–.28

TOLERABLE MISSTATEMENT
- Allowance for sampling risk .... App. D
- Assertions in sample design 4.59
- Classical variables sampling 7.16,
  .... 7.20–.21, 7.27–.31, 7.38, 7.41, App. D
- Defined 4.48
- Expected amounts in 4.60–.61
- Gross margins in sample planning 4.58

TOLERABLE MISSTATEMENT—continued
- Monetary unit sampling ..... 6.24, 6.30, 6.32, ... 6.49, 6.53, 6.56–.57
- Nonstatistical sampling case study 5.05, 5.08–.09, 5.14
- Performance materiality .... 4.50–.59 ....(table 4-3)
- Projected misstatement to population 4.79
- Reclassifications 4.57
- Substantive tests of details .... 4.11, 4.26, ... 4.48–.59 (table 4-3), 4.72
- Test level sampling risk 4.93–.100

TOLERABLE RATE OF DEVIATION
- Determination of 3.48–.54
- Expected population deviation rate .... 3.56–.57 (table 3-3)
- Population size .... 3.59–.61 (table 3-4)
- Sample risk evaluation 3.76–.79
- Sample size and .... 3.41 (table 3-1), ... 3.52 (table 3-2)
- Sampling procedure documentation 3.96–.98
- Sequential sampling tables .... App. B
- Tests of controls .... 3.42–.43, 3.45–.46, ...3.48–.56, 3.76–.79

TRANSACTION PROCESSES
- Gross margins, sample planning and 4.58
- Nonstatistical sampling case study 5.10
- Substantive tests of details .... 4.02, 4.07, ...4.14, 4.58
- Tests of controls 3.14, 3.16, 3.25–.28
- Walkthroughs of 1.07, 3.25–.28

TWO-SIDED RISK PROJECTION, CLASSICAL VARIABLES SAMPLING 7.21

TYPE I MISSTATEMENT. See incorrect rejection, risk of

TYPE II MISSTATEMENTS. See incorrect acceptance, risk of

U

UNAPPLIED CREDITS, MONETARY UNIT SAMPLING 6.08

UNDERLYING DATA TESTING, CLASSICAL VARIABLES SAMPLING 7.06

UNDERRELIANCE, RISK OF
- Audit sampling characteristics 1.28
- Monetary unit sampling 6.03, 6.07, 6.33
- Substantive tests of controls 4.47
- Tests of controls and 3.38–.41

UNDERREPRESENTATION OF POPULATIONS, CLASSICAL VARIABLES SAMPLING 7.31, 7.35–.39

UNDERSTATED BALANCES
- Classical variables sampling .... 7.05–.06, ... 7.42

UNDERSTATED BALANCES—continued
- Monetary unit sampling ...... 6.03, 6.07–.08, ... 6.25, 6.32–.34, 6.35–.37, ... 6.48, 6.53, 6.55, 6.59, 6.63

UNPROJECTED MISSTATEMENTS
- Monetary unit sampling .... 6.23–.25 ....(table 6-1)
- Substantive tests of details .... 4.95, ... 4.99–.102, 4.102–.104

UNSTRATIFIED ATTRIBUTES SAMPLING
- Classical variables sampling 7.12, 7.17
- Process 2.35
- Substantive tests of details 4.28

UNTESTED BALANCES, AUDIT SAMPLING EXCLUSION OF 1.18

UNUSED/INAPPLICABLE DOCUMENTS 3.68, 4.76

UPPER-LIMIT APPROACH
- Actual rate of deviation 3.54
- Attributes statistical sampling tables ... App. A
- Classical variables sampling 7.27
- Monetary unit sampling, 100 percent misstatements .... 6.36–.41 (table 6-2), ... 6.44–.51, 6.63
- Tolerable rate of deviation 3.53

USAGE ESTIMATIONS, TESTS OF CONTROLS 3.16

V

VALUATION AMOUNTS, CLASSICAL VARIABLES SAMPLING 7.06

VARIABLES SAMPLING. See classical variables sampling

VARIATION IN POPULATION
- Classical variables sampling 7.17–.24
- Nonstatistical sampling case study 5.08
- Substantive tests of details 4.27–.32

VOIDED DOCUMENTS
- Substantive tests of details 4.76
- Tests of controls 3.67

W

WALKTHROUGH PROCEDURES
- Audit sampling characteristics 1.07
- Tests of controls 3.25–.28

WEIGHTED SELECTION, SUBSTANTIVE TESTS OF DETAILS 4.21–.22

"WORST CASE" SCENARIO, MONETARY UNIT SAMPLING 6.30

Z

ZERO BALANCES 4.07, 6.07, 7.04

ZERO SAMPLING RISK, CLASSICAL VARIABLES SAMPLING 7.12–.15

AICPA® Online Professional Library

Powerful Online Research Tools

The AICPA Online Professional Library offers the most current access to comprehensive accounting and auditing literature, as well business and practice management information, combined with the power and speed of the Web. Through your online subscription, you'll get:

- Cross-references within and between titles — smart links give you quick access to related information and relevant materials
- First available updates — no other research tool offers access to new AICPA standards and conforming changes more quickly, guaranteeing that you are always current with all of the authoritative guidance!
- Robust search engine — helps you narrow down your research to find your results quickly
- And much more…

Choose from two comprehensive libraries or select only the titles you need!

With the *Essential A&A Research Collection*, you gain access to the following:
- AICPA Professional Standards
- AICPA Technical Practice Aids
- PCAOB Standards & Related Rules
- Accounting Trends & Techniques
- All current AICPA Audit and Accounting Guides
- All current Audit Risk Alerts

One-year individual online subscription
Item # ORS-XX

OR

Premium A&A Research Collection **and get everything from the *Essential A&A Research Collection* plus:**
- AICPA Audit & Accounting Manual
- All current Checklists & Illustrative Financial Statements
- eXacct: Financial Reporting Tools & Techniques
- IFRS Accounting Trends & Techniques

One-year individual online subscription
Item # WAL-BY

You can also add the FASB *Accounting Standards Codification*™ and the GASB Library to either collection.

Take advantage of a 30-day free trial!
See for yourself how these powerful online libraries can improve your productivity and simplify your accounting research.

Visit **cpa2biz.com/library** for details or to subscribe.